"Ted Nunez roars through technologically-accelerating American culture—where it's tough to tell progress from regress from destruction—with a smarter Catholic Social Thought in his soul and *Laudato Si* in his back pocket. Seeking 'sustainable abundance' where others see dystopia, he delivers a sharp analysis of our risky, runaway society and, with a tango-spiced eco-social theology, distills new practices of Catholic social wisdom that should guide any serious discussion about the future of humanity and the planet."

—Anthony J. Godzieba, Villanova University

"Nunez offers us an informed, elegant, and prophetic vision for a future of sustainable abundance. Rooted in the tradition of Catholic social principles and aimed at the world-risking technocracy emerging around us today, this book evocatively invites us to the 'adjacent possible,' a future we confront not with despair but along axes of grace, joy, realism, and hope. A wonderful read, this book is an essential prod to creatively re-imagine our economic and ecological tomorrow."

—James Calvin Davis, author of
  Forbearance: A Theological Ethic for a Disagreeable Church

Sustainable Abundance for All

Sustainable Abundance for All

Catholic Social Thought and Action
in a Risky, Runaway World

TED NUNEZ

CASCADE *Books* • Eugene, Oregon

SUSTAINABLE ABUNDANCE FOR ALL
Catholic Social Thought and Action in a Risky, Runaway World

Copyright © 2018 Ted Nunez. All rights reserved. Except for brief quotations in critical publications or reviews, no part of this book may be reproduced in any manner without prior written permission from the publisher. Write: Permissions, Wipf and Stock Publishers, 199 W. 8th Ave., Suite 3, Eugene, OR 97401.

Cascade Books
An Imprint of Wipf and Stock Publishers
199 W. 8th Ave., Suite 3
Eugene, OR 97401

www.wipfandstock.com

PAPERBACK ISBN: 978-1-5326-1379-1
HARDCOVER ISBN: 978-1-5326-1381-4
EBOOK ISBN: 978-1-5326-1380-7

Cataloguing-in-Publication data:

Names: Nunez, Ted.

Title: Sustainable abundance for all: Catholic social thought and action in a risky, runaway world / Ted Nunez.

Description: Eugene, OR: Cascade Books, 2018 | Includes bibliographical references and index.

Identifiers: ISBN 978-1-5326-1379-1 (paperback) | ISBN 978-1-5326-1381-4 (hardcover) | ISBN 978-1-5326-1380-7 (ebook)

Subjects: LCSH: Human ecology—Theology. | Technology—Religious aspects—Christianity. | Religion and science. | Title.

Classification: BL240.3 .N69 2018 (print) | BL240 (ebook)

Manufactured in the U.S.A.   DECEMBER 21, 2017

Table of Contents

Acknowledgments vii
Foreword ix

Chap. 1 - Introduction: Where the Eco-Social Question Takes Us 1

PART I – *The Sustainable Development of Catholic Social Teaching*

Chap. 2   World Risk Society and the Contested Meaning
of Sustainable Development 15
Chap. 3   The Church's Ambivalent Response to Techno-Economic
Progress (1891–1981) 42
Chap. 4   Church Teaching on Sustainable Development
and Technology (1987–2004) 68
Chap. 5   The Logic of Gift and Integral Ecology 89

PART II – *Christian Praxis and Moral Leadership
in a Risky, Runaway World*

Chap. 6   Freedom, Hope, and the Practice of Envisioning
the Adjacent Possible 115
Chap. 7   Pursuing the Adjacent Possible of Sustainable Abundance 135
Chap. 8   Good Lives, Automation, and the Transition to
Post-Jobs Society 159
Chap. 9   Technology, Up-Down Politics, and the Creative Middle 188
Chap. 10   Beyond the Device Paradigm:
A Theo-Etiquette of Tango 215

Bibliography 237

Index 249

Acknowledgments

Living in D.C. during the scorching summer of 1988—the year both Congress and the country began to take notice of global warming—I met a young Colombian biologist named Jesús, who was in town on a research grant at the Smithsonian. A quiet and thoughtful guy, Jesús was interested in the least among us, the countless little critters and flowers of Amazonia. He was preparing for a multi-year research project set up by Conservation International to identify and study as many species as possible "before it's too late," he said. I vividly remember my impulse to denial upon hearing those words. This book has its beginnings in that moment.

While studying at The Catholic University of America, Dan Cowdin introduced me to the work of Holmes Rolston, who later became a gracious mentor when I wrote my doctoral dissertation on his environmental philosophy. I also benefited immensely from the teaching and insights of several Lonergan scholars, notably Bill Lowe and Cindy Crysdale, while Ken Himes showed me why the Catholic social tradition is worthy of careful study. Readers of Joe Holland's *Modern Catholic Social Teaching* and Mary Hobgood's *Catholic Social Teaching and Economic Theory* will recognize their influence on my thinking.

I am indebted to the members of the College Theology Society, many of whom have provided helpful feedback on my research over the years. I owe a special debt to Jame Schaefer, Phil Sakimoto, Christine Fletcher, James Calvin Davis, and Tony Godzieba—all of whom read and commented on parts of the manuscript. During the 2002–2003 academic year, a visiting appointment at Middlebury College exposed me to the best thinking in environmental studies and the ferocious energy of Bill McKibben, whose leadership of 350.org is more important than ever now that climate deniers control the White House. A summer research grant from Georgian Court University in 2006 afforded me the opportunity to investigate the aftermath of Hurricane Katrina and to meet with and learn from environmental-justice activists. One aim of *Sustainable Abundance for All* is to explain and defend the strategies and actions of eco-justice groups on the front lines in southern Louisiana, North Dakota, and elsewhere.

I've gained much from discussing the ideas of basic income and post-jobs society with Justin Letizia, Robert Braun, and many other bright students at Holy Ghost Prep in Bensalem, Pennsylvania. I am grateful to Kelly Ray and the late Lesley Mitchell, my first tango instructors. When it comes to dancing well in close embrace with each other, my best teacher by far has been Shubhra Wells. For lessons unlearned and dues unpaid, I take full responsibility.

The present work is dedicated to Jess, Llewie, Prem, Priyanka, and Prasad. May the idea and ideal of sustainable abundance for all inspire you as it has inspired me.

Foreword: Christian Futurism

There exists, nowadays, such a thing as a futurist. Some people even do it as their job—attempting to attach some rigor and usefulness to making statements about the Undiscovered Country of things to come. In downtown Palo Alto, the heart of Silicon Valley, there's a storefront for the Institute for the Future. I've attended meetings there; it's for real. Inside, people actually think and talk and write about the future—with, if I may say so, a pretty respectable amount of rigor and usefulness.

Before there were futurists, there were prophets. The Hebrew prophets, for instance, frequently made statements about the future, which could be detailed. But their purpose was not quite that of the futurists. While also examining tendencies in the world around them and identifying disruptive changes in the works, they heard and echoed the voice of God. Their concern was more moral than utilitarian. I'm far from alone in believing that today, with moral hazards around every world-historical corner, we need prophets. But they also need to be futurists. The future is more proximate than ever, ever more an incipient feature of the present.

Christianity has had a peculiar relationship with prophets, and with the future. Many of the faith's early truth claims rest on the fulfillment of the Hebrew prophets' predictions, but Christians' own predictive track records have been less impressive. The longstanding (and ongoing) practice of picking dates for the end of the world has achieved, at least so far, a zero-percent success rate. There is a difference in kind, too, between predicting future events within history and predicting the end of history altogether; to the extent that Christians have been interested in things to come, it has been overwhelmingly on the side of the latter. And this is a tendency we come by honestly. St. Paul encouraged concern with apocalypse over concern for the future, and Jesus suggested that we regard the future as if we were lilies growing in a field. To do otherwise, to attempt to control the future, could be a symptom of pride, a lack of faith.

Many of the reigning ideologies of the twenty-first century adopt this symptom as a cure—hence, the futurists. They tell us that we must have our attention trained on the future, at risk of falling behind. We'll adopt the latest gadgets too slowly; we'll fail to invest in the next disruptive unicorn.

Some Christians refuse to play a part in the contest among futures. They shun new devices or neglect mounting ecological disasters. They prefer the comforts of an imagined past. But what they don't recognize is that the supposed futures are actually features of the present we now inhabit—only disguised as futures in order to discourage human agency in the face of them. While scrambling to keep up with the future, we don't bother to question the assumptions of the here-and-now, or to act.

Climate change, for instance, is not the future but the present, displacing millions of people around the world already. Automation of human jobs by robots and artificial intelligence is happening now, and the prospect of more is causing billions of dollars to flow to the owners of ascendant tech monopolies. A return to feudal levels of economic inequality is well underway. And the worst dystopian fantasies are already here, as we cheerfully funnel the data of our lives and relationships to corporations in exchange for convenient cloud-based services. Speaking of these things as mere possible futures is how the principalities that pretend to rule this world convince the rest of us to wait and see, to treat history as a spectator sport. And lots of Christians are happy to oblige. We keep ourselves busy dwelling on the subset of the present that we imagine to be the past.

What would a prophet do? People like to think of prophets as predicting the future, because predicting is an entertaining game, and gamification dampens the prophets' moral seriousness. But when we study them more closely, we find that's not quite right. More than predicting the future, prophets specialize in perceiving the voice of God in the present—in light of the fact that there is a future in the first place, and that we in the present have some holy responsibility to be co-creators of it. John the Baptist, the last of the prophets in Christian eyes, summed up the overall message of prophecy nicely: *Repent!* Act now, before it is too late. Be accountable.

A prophetic posture today means unveiling the features of the present that come disguised as mere possible futures, participating in the world they've created but refusing to acknowledge the hegemony of any power other than the God who is love. It matters little that the New York Catholic Worker refuses to have a Facebook page (although its founder, Dorothy Day, used whatever publication technologies were available to her in her time), or that my parish has a quite vibrant one. Christianity is not a religion of purity laws. It's a faith that inclines us to ask, instead, how we would redesign Facebook for people truly made in the image of God. What if we did? Can we? How about now? Christian futurists can't just notice and predict technological trends, and leave it at that, because we recognize technology as an ongoing act of co-creation of the universe with our own Creator. We are implicated.

Ted Nunez is among the very few Christian futurists working today, period, and he practices that craft prophetically. Rather than retreating to some imagined past, he employs the tradition of Catholic social teaching in order to see the present, and its supposed futures, with clearer eyes. In the pages to come, expect to find unflinching treatments of topics that are otherwise nearly nonexistent in theological literature but are of immense theological concern, such as technological unemployment, artificial intelligence, self-driving cars, distributed energy production, participatory budgeting, the Internet of things, 3D printing, commons-based licenses, universal basic income, benefit corporations, Moore's Law, the Singularity, and tango. Over and over, Nunez confounds tech culture's claim that the new has nothing to learn from the world from which it came. But he is also at home in tech culture and sees the opportunities it presents for a more equitable and habitable commonwealth to come.

The surest sign of a living tradition and a living faith is when that tradition and that faith serve not merely as places of refuge but as ramparts. They allow us to see the world as it is, without fear or anxiety, and to participate in its tumult without mistaking its totems for our God. The future—the real future—is Creation at work. It awaits our participation.

—Nathan Schneider

1

Introduction:
Where the Eco-Social Question Takes Us

> It has become a commonplace recently to say that we are in a situation where the end of the world is now imaginable—but the end of capitalism isn't.
>
> —BRIAN MASSUMI

> Humanity is now stuck with a planetary stewardship role. We are as gods and *have* to get good at it.
>
> —STEWART BRAND

> It is not a finished Utopia that we ought to desire, but a world where imagination and hope are alive and active.
>
> —BERTRAND RUSSELL

The Great Acceleration

Las Vegas falling quickly behind us, Manjul jacks the rented BMW X5 up to 95 mph as we race past Lake Mead on Route 93. Sanjay rides shotgun, relegating me to the back seat. These bright, successful men from Silicon Valley find the SAV's giddy-up more intriguing than the "bathtub ring" of calcium carbonate rimming the reservoir's shoreline. I interrupt their chatter about the virtues of V-6 engines by pointing out the hundred-foot drop in water level. I can't resist recommending *The Water Knife* for a glimpse into how ugly things could get in the Southwest if the climate-science projections come to pass.

Manjul juices the Beamer into triple digits as he breezily assures me that advances in desalination, irrigation, and water recycling will solve the problem. Sanjay chimes in that his relative Manoj, a billionaire philanthropist, has developed an affordable desalination device called the Rain Maker. Stack a few thousand of them up on a hundred or so barges anchored five miles offshore, and L.A. can release its privileged claim on the Colorado. A bit nonplussed, I redirect the conversation to the Great Pacific Garbage Patch, an inconvenient truth twice the size of Texas floating off the coast my comrades inhabit. The two smile in tandem. Seconds later, Sanjay hands me his Android, and I slump back into the cushion to watch a YouTube video on the Dutch wunderkind Boyan Slat and his Ocean Cleanup Array.

Who knew we were getting so good at saving the planet!

Sanjay's vision soars far higher than barges and floating booms for processing an ocean of plastic. We'll be off to other planets sooner than people realize, he opines. What Musk is doing with Space X is just the beginning. Before I know it, he has me watching a clip of the Falcon 9 rocket landing at Cape Canaveral. This prompts Manjul, who works at Netflix, to mention that *Planet of the Apes* was filmed up at Lake Powell. When I suggest we turn around and visit the Nevada Test Site instead, neither is amused. How about Burning Man, then? That, at least, draws a smirk.

We're rocketing along now at 106 mph as we cross the Hoover Dam Bridge. I ask Manjul if he might want to dial it back just a tad, seeing as how we just passed a "SPEED LIMIT ENFORCED BY AIRCRAFT" sign and all. Both of them laugh. For good measure, Sanjay shows me one more miraculous video, this one a promo for Terrafugia's TF-X. Ah, the flying car every Jetson desires! I'd almost lost hope.

A mild vertigo sets in at 111 mph. Too much screen time? Perhaps sunset at the Grand Canyon will bring us to our senses and provide some perspective on this breathless existence. With Manjul at the wheel, it won't take long to get there.

Building huge dams and detonating atomic bombs in the desert seemed like a good idea to many in the 1950s—bold declarations of freedom from floods, hunger, and tyranny. Mechanized agriculture and precision manufacturing, nuclear energy and synthetic chemicals, cybernetics and television, jet-powered aircraft and astronautics, new vaccines—all were fast developing at mid-century. In 1949, Cardinal Emmanuel Suhard of Paris observed how "modern inventions produced with increasing rapidity cannot be for Christians just another news item or a mere scientific curiosity . . . for they are the making of a new universe. And this is the universe we are called upon to save."[1] The cardinal's reflection on the *res nova* of his day seems prophetic in light of emerging technologies: robotics/AI, synthetic biology, nano-materials, 3D printers, and the Internet of Things, we are told, are poised to disrupt one industry after another. As digitalization, machine learning, CRISPR, and other innovations scale up, productivity will soar. Yet accelerating automation also seems likely to generate higher levels of structural unemployment.

Consider the fate of workers in the living-wage movement. As fast-food employees fight for $15, a burger-bot built by Momentum Machines cranks out 360 gourmet hamburgers an hour. The San Francisco start-up is opening "smart restaurants" and marketing their invention to major players in the industry. Meanwhile, customer-facing automation is working well for Eatsa, which is opening new lunch spots in major cities across the country. Projected cost savings are considerable as is the expected impact on a 3.6 million fast-food workforce.

A similar dynamic is at work in other industries, both in the US and globally.[2] The iPhone6 in my coat pocket is made by Foxconn, which is replacing its million-plus Chinese workers with robots. According to the World Bank, the "share of occupations that could experience significant automation is actually higher in developing countries than in more advanced ones, where many of these jobs have already disappeared."[3] About two-thirds of all jobs in the global South are vulnerable to automation in coming decades.

Both fast food and its automation exemplify what Pope Francis in his encyclical *Laudato si'* refers to as "rapidification" (LS #18): the relentless process of social speed-up driven by technological innovation for the sake of profit. Since 1950 the increasing power, speed, and efficiency of technology has made possible what some scholars call the "Great Acceleration."[4] Population and world GDP, damming of rivers and water use, fertilizer and paper consumption, foreign direct investment and fast-food outlets, telephones and motor vehicles, international travel and carbon-dioxide emissions—all these and much more have increased dramatically from 1950 onward, so much so that 7.4 billion humans now generate a $75 trillion

---

1. Suhard, "Priest in the City," quoted in Haigerty, *Pius XII and Technology*, xix.
2. Ford, *Rise of the Robots*, 12–15.
3. World Bank, *World Development Report* (2016), http://www.worldbank.org/en/publication/wdr2016.
4. McNeill and Engelke, *Great Acceleration*.

world GDP (2016) and at latest tally consume about a quarter of the earth's net primary production.

For the well-educated wizards soaring about, what does it mean to be grateful for all this abundance? And what about the mass of Muggles, growing numbers of whom deeply resent being left behind?

The Day before Thanksgiving

Coupled closely to the technological logic of efficiency are two other logics driving the Great Acceleration forward: the economic logic of capital accumulation and social logic of status-driven consumption. These entwined logics are sanctioned in turn by cultural myths that continue to shape collective identity and shared conceptions of the good life (even as rapid technological and social change leaves many in "present shock"[5] and hastens the collapse of narrative meaning).

As an example, consider the two-part editorial titled "The Desolate Wilderness . . . And the Fair Land," which has appeared on the day before Thanksgiving in the *Wall Street Journal* every year since 1961. The first part is not an editorial, but rather a well-known passage from colonial history: Governor Bradford's account of the pilgrims' tearful departure from Delfs-Haven and harrowing first contact with a "hideous and desolate wilderness." The second, written at the apex of the American Century, seeks to renew the colonial taming of wilderness quest "in a time of troubles." It opens with an admiring survey of what the Protestant work ethic has wrought across the continent since Plymouth landing. Beholding a "fair land," the traveler-narrator imagines still greater days ahead: "America, though many know it not, is one of the great underdeveloped countries of the world; what it reaches for exceeds by far what it has grasped."[6] Yet the mood of the country is quite different and troubling questions cannot be dismissed: America the bountiful and free is wracked by "social discord" within and under threat from "unpredictable strangers" without. Read a half-century later, it sounds all too familiar.

How can the country move forward with hope and courage? By reminding ourselves, the traveler-narrator intones, "that the richness of this country was not born in the resources of the earth, though they be plentiful, but in the men that took its measure." By reminding ourselves "that for all our social discord we yet remain the longest enduring society of free men

---

5. Rushkoff, *Present Shock*.
6. Editorial, "Desolate Wilderness . . . Fair Land," *Wall Street Journal*, November 25, 2015.

... the marvel and the mystery of the world, for that enduring liberty is no less a blessing than the abundance of the earth."[7] These paired texts annually reaffirm the US business elite's self-image as pioneers of progress and keepers of the entrepreneurial spirit. Under *WSJ*'s neoliberal banner, the old manifest-destiny message is re-presented with hardly a blush as the best way forward even now.[8] Among the affluent, its Promethean-cornucopian imagination resonates deeply.

Propelled not only by technology and capital but also by powerful storylines and the carefully calibrated economy of desire and social position, the Great Acceleration has hurtled humanity into the Anthropocene, a new geological epoch in which "human activities have become so pervasive and profound that they rival the great forces of Nature and are pushing the Earth into planetary *terra incognita*."[9] As a result, conservation biologists estimate that a third to half of the planet's species will vanish before the close of the current century. Given the Great Acceleration's immense momentum and cumulative, long-term impacts on planetary life systems, it is an open question whether a transition to a just, humane, and sustainable global society—one in which both humans and myriad other species may flourish—is possible. Call it the eco-social question, the compelling question of our collective future.

A Moment near Midnight

The Great Acceleration can be correlated with the rise of what sociologist Ulrich Beck calls "world risk society," a world in which technological and environmental mega-hazards call into question taken-for-granted conceptions of progress.[10] Oppenheimer's recollection of Lord Krishna's words—"Now I am become death, the destroyer of worlds"—captures the sense of foreboding that has become perpetual since the success of "Trinity" (the

7. Ibid.

8. Critical theorists are quick to note how the *WSJ*'s narrative of colonial conquest unwittingly reinforces the ongoing erasure of the threatening, wild other in the name of bringing law to the land in the Americas and other presumably "backward" regions. For 500 years now, a deadly logic of the One and the Other has authorized the brutal exercise of colonial power from the Arctic to Patagonia; see, e.g., Roth, *Annihilating Difference*. Liberation and feminist theologians add that biblical narratives cannot be divorced from the history of their effects. Exodus sets up the "texts of terror" that follow (e.g., Josh 6–12); taking possession of promised lands and eliminating enemies in the process is what the faithful are called to do, then and now. See, e.g., Hawk, "Truth about Conquest"; and Wright, *Myths America Lives by*.

9. Steffen et al., "Anthropocene," 614.

10. Beck, *World Risk Society*.

ironic code name for the atomic detonation at Los Alamos in July 1945) and its real-world application on the civilian populations of Hiroshima and Nagasaki.

First set by atomic scientists in 1947 at seven minutes to midnight, the Doomsday Clock came to symbolize the MAD logic of geopolitics and perpetual threat of nuclear annihilation during the Cold War era. After the Wall crumbled the iconic clock was dialed back to seventeen minutes in 1991, and some thought its end—along with history—was at hand. But the scientists had reason to let the Clock of Doom keep ticking. The big hand was nudged forward in 2015 from five to three minutes before midnight because of growing concern over the possibility of runaway climate change, existential risks related to emerging technologies, and plans among the powerful to modernize nuclear arsenals.[11] During inauguration week in January 2017 the big hand was moved thirty seconds closer.

While many continue to deny that we live in a moment near midnight, some among us sense it acutely and refuse to remain quiet. Indeed, today's world risk society possesses a hydra-like quality. Environmental protests seem to pop up continually with aggrieved publics demanding accountability from corporations and governments. As soon as one campaign ends, two others begin. Anti-GMO protesters gather outside Starbucks to demand transparency. In the Adirondacks, activists prod small towns to enact bans on fracking. Thousands hit the streets in cities across the US to oppose the Keystone XL Pipeline, while hundreds of first-nation activists gather at Standing Rock to stop the Dakota Access Pipeline. Kayaktivists form a flotilla underneath hang-activists suspended beneath the St. Johns Bridge, temporarily blocking a Shell Oil vessel's departure from Portland on its way to an Arctic drilling site. Students on dozens of campuses push for hydro-carbon divestment. A "People's Climate March" in NYC draws half a million accompanied by 2,600 solidarity events in 162 countries. As sociologist Barbara Adam observes, these and hundreds of other protests across the globe reflect an underlying conflict between "temporalities that operate according to different principles, to the variable, rhythmic temporality of nature and the cosmos on one hand and the industrial times of the machine, the laboratory and economic considerations on the other."[12]

A risky, runaway world puts a panicked politics of time into motion. Recently, for example, a group of climate scientists, policy-makers, and corporate leaders issued a post-Paris climate accord report and urgent call to collective action titled "2020: The Climate Turning Point."[13] With a global

---

11. On the increasing danger of nuclear catastrophe, see Perry, *Nuclear Brink*.
12. Adam, *Timescapes of Modernity*, 106.
13. Figueres et al., "Three years."

"budget" of just 600 billion tons of carbon dioxide left to "spend" before humanity exceeds the Paris limit of 1.5 to 2 degrees Celsius, and with the current burn rate at 41 billion tons, the global "we" has but fifteen years before going into climate "debt"—a debt that could prove crushing when "variable interest rates" (i.e., the amplifying effects of various climate feedback loops) kick in. But given the decade or so it will take to turn the global Enterprise on to a path of deep de-carbonization, the Mission 2020 group says we now have only *three more years* to effect far-reaching changes in practices and policies across the planet. Any delay raises the danger future generations will face. In other words, to keep climate change under the speed limit, as it were, societies must vastly *accelerate* the transition to a post-carbon, cleantech economy. The Mission 2020 hashtag? #DontBeLate.

In world risk society, popular consciousness and public decision-making are colored by anticipations of large-scale catastrophe under conditions of volatility, uncertainty, and complexity. Elites and ordinary citizens alike are more aware of a globalized thrown-togetherness, an involuntary state of belonging to risk communities whose fate remains caught in perpetual suspense. It's an unsettling situation in which no one knows where or when the next "black swan event"[14] will occur. At the same time, the vulnerability of downstream, downwind, and sea-level communities—some already designated as sacrifice zones—remind everyone that technological and environmental risks ramify not only in unpredictable ways across geographies and generations, but also along familiar lines of class, race, gender, and marginalized culture. The power to define and distribute "bads" has become as salient a political issue as control over the production and distribution of goods.

Beck, Adam, and other risk-society theorists contend that techno-economic progress has become self-endangering, thus creating an ongoing legitimation crisis for governments and corporations and opening up opportunities for social movements and civil-society actors to articulate and act on alternative visions. However, social acceleration not only makes daily life more stressful and time-starved for many but also undermines the prospects for deliberative democracy and collective action. How do we envision and plan for the future when so much changes so often now? Under the "tyranny of the moment,"[15] pressure mounts for decisions to be made quickly and with only short-term consequences in view. Rapid technological and economic developments tend to leave democratic decision-making and long-term policy implementation in the dust. Social acceleration tilts

---

14. Taleb, *Black Swan*.
15. Eriksen, *Tyranny of the Moment*.

the board toward a neoliberal agenda of regulatory rollback, fast technology adoption, and flexible economic arrangements.

The Eco-Social Question

The social question addressed by Pope Leo XIII in *Rerum novarum* (1891) has morphed into a more complex eco-social question, which might be formulated along ethical-practical lines as follows: What constitutes the good life, and how can persons and communities sustain hope for and move responsibly and creatively toward human and planetary flourishing in a risky, runaway world? This study of Catholic social thought and action takes up certain aspects of the eco-social question by exploring the idea and ideal of *sustainable abundance for all*. The multi-disciplinary approach I take invites many viewpoints to the table. At the outset, it may help to indicate some of the perspectives informing the discussions to follow. Both narrative and practice play a role in the dance toward understanding more deeply how sustainable abundance for all might come to fruition.

From an eco-theological perspective, sustainable abundance for all gestures first and foremost toward the experience of grace, understood here as a deeper and more intimate encounter with unfolding plenitude in universal history. God creates and sustains a mysterious, evolving cosmos in which humans enjoy an "original blessing"[16] and are invited to participate responsibly and lovingly in the convivial co-creation of greater value and beauty. As Saint Francis has taught us, the abundant life our hearts desire most is received joyfully from a loving, liberating God who calls us into communion with all beings. Out of gratitude and in humility we go forth to embrace all in the dance of life, our lives a work of dramatic artistry dedicated to healing and creating in history.

From the perspective of sustainability science, our home planet's history may be characterized broadly by sustainable abundance for all, having evolved more diverse, complex, and sentient lifeforms over 3.5 billion years to bequeath the present generation with perhaps eight-million-plus fellow species (many of which remain unknown and unnamed, the unfinished work of Adam). However, the remarkable story of biological and cultural evolution has taken an ominous turn of late. The "relatively stable, 11,700-year-long Holocene epoch, the only state of the planet that we know for certain can support contemporary human societies, is now being destabilized."[17] Scientists warn that global society's transgression of

16. Fox, *Original Blessing*.
17. Steffen et al., "Planetary boundaries."

planetary boundaries—and especially the two core boundaries of climate change and biosphere integrity—may "inadvertently drive the Earth system to a much less hospitable state."[18] These systemic risks appear heightened to the extent that high-speed capitalist society appears incapable of adopting a slower, precautionary approach to techno-economic development. With the ecological economists, I contend that societies can achieve sustainable abundance for all—i.e., the flourishing of both humanity and myriad other species on the home planet—only by renouncing the dangerous dogma of economic growth and constructing dynamic, steady-state economies within bounded communities.[19]

In a risky, runaway world the practice of envisioning and pursuing the adjacent possible of sustainable abundance for all places a premium on empathy and imagination and moves us into ongoing debates over approaches to sustainable development, social-change strategies, and democratic processes for managing technological and environmental risks. Grounded in an *evolutionary ethics of responsible care*, the pursuit of sustainable abundance for all takes political shape through a new-distributist program for building inclusive, prosperous, and just communities that achieve a situated fitness within their respective local-regional ecosystems and the biosphere. Finally, sustainable abundance for all is embodied and lived out through community-based practices such as participatory democracy, citizen science, cooperative enterprise, co-housing, social dancing, and liturgy. These and other practices flourish beyond the treadmill of mindless labor and distracting consumption, or what philosopher Albert Borgmann calls the "device paradigm."[20] They enable us to engage in meaningful work and enjoy fruitful leisure—both of which are antidotes to the alienations of a technological society caught within what social theorist Hartmut Rosa refers to as the "circle of acceleration."[21]

Dancing in Close Embrace along the Cliff's Edge

John Coleman, SJ once said every theology presupposes a sociology and vice versa. My aim is to bring Catholic social thought into dialogue with contemporary social theory and philosophy of technology for the purpose of fashioning a constructive and hopefully artful Christian response to certain challenges facing our world, in particular the search for pathways

18. Ibid.
19. Daly and Cobb, *For the Common Good*.
20. Borgmann, *Technology*.
21. Rosa, *Social Acceleration*.

toward sustainable development and impact of industrial and emerging technologies on culture and society. In *Laudato*, Francis critiques the dominant technocratic paradigm, presents a broad vision of integral ecology, and calls for ecological conversion and a bold cultural revolution. Taking up these themes, I argue that the emergence of a world risk society driven by a self-propelling process of social acceleration has created a compelling need for new forms of Christian praxis and new kinds of Christian leaders. Like street dancers collaborating at the cutting edge, we need to invent and experiment with new steps and figures, new methods and movements at once innovative and grounded in traditions of Christian social wisdom, civic engagement, and cultural renewal.

Part I provides a fresh reading of modern Catholic social teaching by focusing in holistic fashion on the nexus of technology, economy, politics, and ecology. I examine how the church has engaged the decades-old sustainable-development debate as well as the emerging proactionary-precautionary debate between techno-progressives and bio-conservatives over technology's trajectories. In tracing out continuities, tensions, and shifts in the church's evolving response to techno-economic progress and its dark side, I attend to the vital role of social Catholicism in shaping papal and episcopal social teaching from 1891 to the present. I conclude that recent Catholic social teaching opens the way for new forms of Christian praxis and moral leadership aimed at building sustainable communities, redesigning institutions, and resynchronizing natural and social ecologies in a range of contexts within privileged as well as impoverished, marginalized locales across the globe.

Taking Part I's conclusion as a promising point of departure, Part II explores possible avenues for advancing the "great work" of societal transformation in the twenty-first century.[22] After clarifying my theological-ethical stance (chapter 6), I develop a future-oriented, strategic response to some of the major political-economic and environmental challenges facing individuals, communities, and nations in world risk society (chapter 7). Among the issues addressed in detail is the other inconvenient truth of technological unemployment and how we might transition to a "post-jobs society" in coming decades (chapter 8). Another is the ongoing debate over which approaches to technology assessment are best adopted under conditions marked by uncertainty and risk (chapter 9). Still another is the cultural danger of becoming "people of the device" who tend to avoid face-to-face encounters, intimate conversations, close embraces, and other forms of connection and commitment necessary not only for individual happiness and

22. Berry, *The Great Work*.

healthy community life but also social change (chapter 10). Underlying all of these discussions is some version of the eco-social question, which invites us to ponder what constitutes the good life and how we can move toward human/planetary flourishing in today's high-speed technological society.[23]

The audience I have in view includes educated, privileged Christians in the US and elsewhere who recognize the need for new models of community, social action, and moral leadership in a time of accelerating technological and social change. Much can be gained from thinking through what the logic of gift (Benedict XVI) and integral ecology (Francis) mean in theory and practice. In particular, how we (re)engage with "focal things and practices" (Borgmann) and how we learn to operate responsibly within multiple "timescapes" (Adam) is central to the ethical-practical challenges before us. To live a sustainably abundant life in a risky, runaway world, we need to take up a *politics of engagement and time*. This kind of moral politics requires both ordained and lay Christian leaders (and leaders from other faith traditions as well) who dare to imagine and make good on their shared visions, who lead with humility even as they experiment boldly, who speak with insight and compassion to contemporary realities such as rapid techno-economic development amidst gross inequality, and who thus offer a genuinely hopeful reply in word and deed to those who claim, in resignation or despair, that no alternative to the juggernaut of technocratic capitalism exists or that time to respond effectively to global environmental crises already has run out.

While the analyses and arguments ahead take the Great Acceleration and ticking Doomsday Clock with full seriousness, none presume we're fated to end the way of *Thelma and Louise*. At the end of Ridley Scott's 1991 road film, Thelma (Geena Davis) and Louise (Susan Sarandon) succumb to despair and meet their Maker after flooring a Ford Thunderbird over the cliff of the Grand Canyon. In this moment near midnight, I propose we attend to deeper rhythms and learn to dance in close embrace along the cliff's edge, clear-eyed about the dangers and ongoing destruction, compassionate toward all we see, and committed to weaving our way artfully and with passion toward a sustainably abundant world for all.

---

23. Following the pioneering work of Albert Fritsch, Roger Shinn, Paul Durbin, Ian Barbour, Philip Hefner, Frederick Ferre, John Hart, and others, second-generation discussions in the subfield of theological ethics and technology are growing. It is not my aim here to present a comprehensive survey or overview of the field at present. Among the related areas of study and current topics not addressed in-depth by this work are media ecology (McLuhan, Postman et al.), cultural environmentalism (Boyle et al.), science and technology studies (Latour et al.), the rise of ubiquitous surveillance, and transhumanism.

Part I

The Sustainable Development of Catholic Social Teaching

2

World Risk Society and the Contested Meaning of Sustainable Development

> Technological change is not additive; it is ecological. A new technology does not merely add something; it changes everything.
>
> —NEIL POSTMAN

> Today we need to slow down and divert human intrusions into various planetary force fields even as we speed up efforts to reconstitute the identities, spiritualities, consumption practices, market faiths and state policies entangled with them. Such a tension helps to constitute the contemporary fragility of things.
>
> —WILLIAM CONNELLY

The social question of a century ago focused on poverty and inequality within industrial society, and today these issues remain as salient as ever. In *Capital in the Twenty-First Century*, economist Thomas Piketty provides a historical analysis of rising inequality in wealth and income. When the rate of economic growth is lower than the rate of return on invested capital, Piketty argues, the share of national income going to owners of capital increases. On this view, that has been the situation since the 1970s and likely will continue in coming decades.[1] At the same time, the world's peoples and their leaders are now faced with a more complex eco-social question within the horizon of world risk society, a world in which

---

1. Piketty, *Capital*.

previously taken-for-granted aspects of modernity—e.g., the unalloyed benefits of technology and economic growth—are problematized. This chapter sets the context for later discussions by surveying the contours of world risk society and dynamics of social acceleration, and by analyzing the contested meaning of sustainable development in public discourse. Along the way I'll raise several questions—and one basic question—for further exploration in the chapters to follow.

A Welter of Manufactured Uncertainties

People in world risk society worry about the latent effects of industrial activities and uncertain impacts of new technologies. While Chernobyl, Bhopal, Exxon Valdez, BSE, the Asian financial crisis (1997–98), 9/11, Katrina, Deepwater Horizon, Fukushima, and other disasters serve as signal events, world risk society arrives for the most part by stealth through the slow proliferation of invisible hazards.[2] Global public concern over the "quiet crisis" of the environment has grown over the decades as problems of pollution and over-population have been joined by ozone depletion, climate change, mass species extinction, collapse of fisheries, deforestation, and freshwater shortages. World risk society emerges when the political question of who gets what and how much unavoidably shifts to who gets exposed (or dumped on) and how much. Disputes over how one defines and distributes various risks (the "bads") force their way into the political scrum over the distribution of goods.[3] Consciousness and decision-making come to focus as much on anticipations of large-scale catastrophe as on ways to keep the wheel of progress rolling.

Ulrich Beck contends we are undergoing a historical transition from early to late modernity, a transition in which "the foundations of the established risk logic are being subverted or suspended."[4] During early modernity (1750 to 1950), industrial society grows confident in its ability to dominate nature, calculate risks, and minimize the impact of industrial hazards by creating technical safeguards, rational methods (e.g., actuarial science), and institutional mechanisms (e.g., insurance) that render industrial risks manageable and provide after-care to cope with accidents. Early "modernity,

---

2. Mention of the late-90s financial crisis in Asia and 9/11 indicate two further dimensions of world risk society—the volatility of global finance and specter of transnational terrorism—theorized by Beck and others. In this study, I focus on technological and environmental risks, though below and in chap. 7 the connections between these risks and the systemic risks associated with global-financial debt are noted.

3. Hajer, *Politics of Environmental Discourse*, 36.

4. Beck, *World Risk Society*, 53.

which brings uncertainty to every niche of existence, finds its counter-principle in a *social compact against industrially produced hazards and damages*, stitched together out of public and private insurance agreements; and, thus, activating and renewing *trust* in corporations and government."[5] For early-modern consciousness, the benefits of new technologies clearly outweigh whatever risks might attend mass adoption. Belief in progress through science remains unquestioned and reaches across ideological, class, and other divides. In this world, industrial risks are *calculable* (and often low), while even the most horrific disasters—a sinking Titanic, an exploding Hindenburg—are *limited* in scope.

According to Beck, the phase of late modernity begins with the emergence of world risk society in the 1940s and grows more acute and complicated with each passing decade. Early-modern assumptions regarding scientific prediction, technocratic control, and risk management still shape the thinking of elites in science, business, and government, but increasingly their risk models, forecasting methods, and monitoring techniques are called into question by counter-experts and ordinary citizens dealing with the latent harms (e.g., cancer clusters) of science-driven progress. Rachel Carson's *Silent Spring* (1962) is emblematic of a more critical, late-modern view of technology's boomerang effects. At first vilified by industry and many in the scientific establishment, Carson's work on pesticide over-use resonated with the public and was vindicated by President Kennedy's Science Advisory Committee. The pattern of corporate denial, obfuscation, personal attacks on researchers, and staging of scientific doubt plays out often in world risk society, with corporate-sponsored denials of global warming the latest episode.[6]

Beck, Barbara Adam, and other risk-society theorists hold that modernity has crossed a threshold with the risks attending nuclear, chemical, genetic, and emerging technologies as well as global environmental problems such as climate change and loss of biodiversity. While early-modern industrial risks were calculable and damage from accidents limited in scope, many of today's technological and environmental risks are more uncertain, complex, and expansive in the scale and potential irreversibility of their impacts. On this view, human domination of nature through

---

5. Ibid., 52.

6. Oreskes and Conway, *Merchants of Doubt*. Among numerous examples, recall DuPont's effort in the early 1980s to organize a lobby group, the Alliance for Responsible CFC Policy, dedicated to countering calls for a ban on chloroflourocarbons (CFCs), which were linked to ozone depletion. The corporation eventually reversed course in 1986 when the science proved indisputable and political support for regulation became all but impossible to reverse.

expanding techno-economic growth becomes self-imperiling. Proliferating mega-hazards threaten to overwhelm early-modern institutional coping capacities. It is telling, Beck observes, that the nuclear power industry is uninsurable and that engineers at Chernobyl and Fukushima were unable to prevent a meltdown or contain the radioactive fallout.[7] Moreover, the scale and complexity of many technologies now makes virtually inevitable the occurrence of what sociologist Charles Perrow calls "normal accidents," since "we are dealing with complex systems in which a series of failures can come together in a way that no one can anticipate."[8]

In world risk society everyone must live with and respond politically to the ever-present "power of threat" (Beck). At the same time, the negative impacts of industrial modernity are unevenly distributed: a world in which all are exposed also produces winners and losers, namely downwind, downstream, and low-lying regions around the world. Poor air quality, lack of potable water, proximity to waste incinerators and heavy-industrial operations, flood-zone occupancy, desertification, soil degradation, lack of access to green space—these and other vulnerabilities affect increasing numbers of marginalized, low-income communities across the planet.[9] The category "environmental refugee" is now tracked by the U.N. and other international agencies. Hence alongside sustainability, the concept of environmental justice has come into wide use among activists in numerous countries. "Emerging from its origins in anti-toxics and civil rights activism in the US to produce what some have seen as one of the most significant developments in contemporary environmentalism, environmental justice has become increasingly used as part of the language of environmental campaigning, political debate, academic research and policy-making around the world."[10] Over the past two decades, environmental-justice movements in the U.S. have moved beyond a local-issue NIMBY mentality to a broader NIABY ("not in anyone's backyard") outlook that finds common cause with eco-justice, peasant solidarity, and indigenous rights movements in the global South. Since 1999, these movements have participated in anti-globalization protests and gatherings such as the World Social Forum, Klimaforum09,

---

7. In Beck's view, the various risks attending nuclear power and waste disposal, climate change, chemical pollution, and other mega-hazards constitute a "systematic violation of basic rights, a crisis of basic rights whose long-term effect in weakening society can hardly be underestimated" (*World Risk Society*, 39).

8. Perrow, *Normal Accidents*, 353.

9. See, e.g., Carmin and Agyeman, *Environmental inequalities*.

10. Walker, *Environmental Justice*, 1; see also Schlosberg, *Defining Environmental Justice*.

and the 2010 People's World Summit on Climate Change and Mother Earth Rights in Cochabamba, Bolivia.

Over time the "manufactured uncertainties" (Giddens) of industrial modernity undermine the claims of experts and institutions to manage risks and promote prosperity. When corporate and governmental actors deny uncertainty and fail to acknowledge the limits of predictive control, it only deepens the distrust felt by a vulnerable populace. Assurances of safety ring hollow in the face of incalculable threats to human and environmental health. Beck refers to the official sanctioning of mega-hazards as the "organized irresponsibility" of modern industrial risk-management practice:

> The unknowable risks of the risk models hide behind the façade of controllability. Since modern forms of risk management for the most part maximize mathematical precision, they systematically underestimate unforeseen and improbable, but not therefore impossible, occurrences, as regards both their frequency and the extent of the damage they cause. This apparently slight difference between "improbable" and "impossible" makes a world of difference.[11]

In a world where inconvenient truths multiply, conflicts over the definition and distribution of risks break out as people question the ability of scientific and legal authorities to determine acceptable levels of safety and manage complex systems. A self-endangering progress generates an ongoing legitimation crisis, which in turn opens up opportunities for social movements and civil-society actors to influence public policy.

The rise of world risk society raises several important questions: In situations characterized by risk, uncertainty, and complexity, what is the most responsible way to develop and assess technologies? Who gets a say in the process, and what ethical principles, conceptual frameworks, and scientific methodologies inform the deliberations? What role do experts and laypersons play in decision-making, and how can they interact meaningfully in a world in which the former typically know "more and more about less and less" (i.e., the problem of specialization), while the latter often feel overwhelmed by the magnitude, complexity, and pace of technological change and incompetent to render an informed judgment on matters "best left to the experts"? These questions are explored at length in chapter 9.

---

11. Beck, *World at Risk*, 130.

Black Swan Blindness

Both policy elites and ordinary citizens are now more aware of a globalized thrown-togetherness—an involuntary state of belonging to risk communities whose fate seems to hang perpetually in the balance. According to risk expert Nassim Taleb, our unsettling situation is this: we know not where or when the next black swan event—the very improbable, high-impact events that shape history—will occur.[12] Among many examples, consider that no one in 1914 saw World War I coming,[13] no one in 1950 (including Carson at the time) predicted silent springs before the end of the decade, no one in 1975 saw Beirut about to explode, no one in 2000 (except Al-Qaeda) saw 9/11 coming, hardly any economists foresaw the financial meltdowns that occurred in 1987, 1997–98, and 2007–08, and no one knows how tame or terrifying climate change will be—the increasing sophistication of climate-science models notwithstanding.

Routine denial of black swan blindness aggravates our predicament further. By banishing black swans from our individual and collective imaginations, we make it easier to throw precaution to the wind and thus make it more likely that the next whirlwind will overwhelm us. Example A is the small but quite real possibility of "climate shock," i.e., a catastrophic scenario *surpassing* the most fearful forecasts of the Intergovernmental Panel on Climate Change.[14] Example B comes from the Great Recession: in October 2008, as broad public awareness of the financial meltdown's full extent took hold, Federal Reserve Chairman Alan Greenspan (usually considered the smartest guy in the room) confessed to Congress that policy assumptions and data inputs to risk-management models had been fundamentally flawed: "I made a mistake in presuming that the self-interests of organizations, specifically banks and others, were such that they were best capable of protecting their own shareholders and their equity . . . The whole intellectual edifice collapsed in the summer of last year because the data inputted into risk management models generally covered only the last two decades, a period of euphoria."[15] As environmentalist Bill McKibben points out, the

---

12. Taleb, *The Black Swan*.

13. Clark, *The Sleepwalkers*. In an admiring review, historian Thomas Laqueur writes: "Clark's story is 'saturated with agency'. Many actors (the crowned heads of Europe, military men, diplomats, politicians and others), each with their own objectives, acting as rationally and irrationally as humans are wont to act, made decisions that foreclosed on others and collectively led the world into an unimaginable and un-imaged war. Collectively, they produced the greatest 'black swan event' in world history" ("Some Damn Foolish Thing").

14. Wagner and Weitzman, *Climate Shock*.

15. Greenspan, quoted in McKibben, *Eaarth*, 105.

gist of Greenspan's mea culpa applies more broadly: "On a larger scale," McKibben writes, "our whole civilization stands on the edge of collapse because the data inputted into our risk management models come from the last couple of hundred years, a very atypical time."[16]

Yet not all of history's black swans are, well, black. Few in 1980 saw the democratic wave rising and the coming fall of one dictatorship after another across the planet. Few in the fall of 2010 predicted an Arab Spring. Crises open up opportunities for social movements and civil-society actors to advance alternative agendas oriented to the adjacent possible of a more just, humane, and sustainable world. As the walls erected by totalizing visions crumble and connectivity becomes ubiquitous, ordinary citizens seize upon new opportunities for movement not simply across geographical borders, but also beyond inherited categories of every kind. Chapters 6 and 7 explore possibilities for and obstacles to social transformation at this moment in history.

You Are on Your Own

Within advanced-industrial countries, the sociological process of individualization has taken hold on a broad scale. Individualization is both a structural reality and a modern cultural ideal shaping the daily experience and life prospects of persons in a risky, runaway world. Loosed from ties to family, community, and tradition and treated by the liberal state as an individual rights-holder, more and more persons aspire to get a life of one's own. Exposed through digital media, education, and sophisticated marketing to a vast array of possibilities, the embedded self gives way to the DIY biography. Identity is formed in more fluid fashion through personal choices in lifestyle and career; aspiration and achievement are gauged by one's ranking within the meritocracy. Stable membership in a large, historic institution (e.g., a union or church) counts for less as social ties increasingly become more voluntary, temporary, and tenuous in character.[17] Under conditions of market liberalization and hyperacceleration, one must move quickly from one shifting situation and momentary experience to the next. Lives thus tend to become episodic, lacking in narrative depth and enduring attachments.

Greater freedom to set one's own course is accompanied by higher levels of risk and insecurity. Since the 1970s, changes in corporate strategy and public policy have aided and abetted hyper-individualization and thus rendered more people "up in the air" rather than grounded. In *The*

16. Ibid.
17. Giddens, *Modernity*; Beck and Beck-Gernsheim, *Individualization*.

Great Risk Shift, political scientist Jacob Hacker details how a combination of management demands for workforce flexibility, elimination of company benefits, and welfare-state retrenchment have left workers to face the vagaries of a volatile labor market and to provide for their own retirement: "Social Security, Medicare, private health insurance, traditional guaranteed pensions—all sent the same reassuring message: someone is watching out for you, *all of us are watching out for you*, when things go bad. Today, the message is starkly different: *You are on your own*."[18] Freedom in the marketplace can mean a freefall into severe economic insecurity and chronic instability, a crisis mode that can be difficult to escape when familial, social, and educational resources are unavailable.

The appearance of Uber, Mechanical Turk, and other "gig economy" organizations is symptomatic of a decades-long trend toward contingent work practices that have fueled a growing sense of insecurity among workers in the US and many other countries.[19] A new global class—the "precariat" (Standing)—made up of blue-, pink-, and white-collar workers relegated to temp/part-time roles, well-educated yet chronically underemployed youth, adjunct faculty and interns, legions of migrant laborers, and a growing number of criminalized people of color now toes the wire with only a thin, fraying net (or dragnet) to catch them when they slip up.[20] Chapter 8 addresses the plight of the precariat within the context of growing inequality and the specter of technological unemployment.

Runaway Technology, Social Acceleration, and the Rise of Up-Down Politics

Part of what makes life in world risk society at once exciting and unnerving is the accelerated pace of technological change. To convey the power of exponential technology, inventor and futurist Ray Kurzweil tells the story, set in ancient India, of the emperor and the inventor of chess, who gifts the game to royalty. Impressed with the smart fellow, the emperor says, "Name your reward." The inventor replies, "I only want enough rice to feed my family, your majesty." Even more impressed, the emperor tells him to name the amount. Looking at the chessboard, the inventor says, "I'll take a dozen large sacks or, if I may, the amount that's finally on the chessboard after the number of rice grains are doubled from one square to the next, until all the squares are filled. So, on square one we'll place one grain, on the next

18. Hacker, *The Great Risk Shift*, x; see also Andolsen, *New Job Contract*.
19. Scheiber, "Rising Economic Insecurity."
20. Standing, *Precariat*.

two grains, then four grains on the third, and so on. Emperor, please, you choose. My family's welfare is in your hands." Smiling, the emperor issues the command, "Fill the chessboard, one square at a time, doubling the number of grains as you go!" At square twelve the emperor is still smiling; there's not enough rice to fill a cup. But once the first half is filled, well over four billion grains have piled up. Before his servants begin to fill the thirty-third square, a now enraged emperor halts the amassing of rice and orders the inventor beheaded.

Accelerating technology development confronts us with the possibility of exponential change and a future shock that would take even Toffler aback. Moore's law—the doubling of computing power every eighteen to twenty-four months—has held true for five decades (though signs of slippage are now appearing). In *The Singularity Is Near*, Kurzweil posits a "law of accelerating returns"—essentially Moore's law generalized—operative in the history of technology.[21] Along these lines, MIT researchers Eric Brynjolfsson and Andrew McAfee argue in *The Second Machine Age* that we are at or very near the second half of the chessboard with the Internet of things, robotics/AI, 3D printing, nanotechnology, synthetic biology, and other emerging technologies poised for take-off.[22]

Concerns about rapid technological change are nothing new, of course. The theme of runaway technology reaches back to the beginnings of modern sociology and threads its way through much of modern thought.[23] Hartmut Rosa has developed a theory of modernization focused on the dynamics of social acceleration and common experience of living in a "runaway world" (Giddens). From the mid-eighteenth century onward, modernizing societies have been caught up in a self-propelling process of technological, social, and cultural acceleration, or what Rosa refers to as the "circle of acceleration" in which *"technical and, above all, technological acceleration serves as a powerful driver of social change."*[24] Each wave of technological innovation generating an industrial revolution increases the pace of social change, which in turn quickens the experienced pace of daily life. With things moving at a faster pace, demand rises for greater labor- and time-saving technology, which then drives a new wave (or cycle) of technological innovation propelled relentlessly toward achieving greater power, efficiency, and speed—up to the lightning-speed of today's information and communications technologies.

21. Kurzweil, *Singularity*, chap. 2.
22. Brynjolfsson and McAfee, *Second Machine Age*, chap. 3.
23. Winner, *Autonomous Technology*.
24. Rosa, *Social Acceleration*, 154.

Rosa sees social acceleration crossing a threshold to late-modern hyperacceleration with the democratic and digital revolutions of the 1980s and 1990s as well as the ascendancy of neoliberal governance during these decades. On this reading, globalization's defining feature is the accelerated pace of flows: "The exchange or movement of information, money, commodities, and people, or even of ideas and diseases, across large distances is not new: what is new is the *speed* and *lack of resistance* with which such processes transpire."[25] Hyperacceleration threatens to erode or "liquefy" modern institutions such as the nation-state, its bureaucracies and military, remaining elements of the Fordist work regime, and even democracy itself—all of which had functioned as accelerators in the phase of "classical modernity" but are now considered impediments to accelerating global flows.

In high-speed society, it becomes imperative to keep up or else pay a severe penalty, e.g., in social exclusion, unemployment, and bankruptcy. Paradoxically, the plethora of life options and contingencies in the current phase of hyperacceleration leads not to greater freedom, but rather to a widespread sense of time scarcity, a compulsion to adapt quickly, and a tendency to fixate on what works in the short run as opposed to making deliberative choices against longer-term horizons of expectation. Under the "tyranny of the moment" (Eriksen), one is "free" to run as hard and fast as possible only to find that one has stayed in the same place, i.e., kept pace (barely) with a fast-changing world. Hyperacceleration in late modernity thus yields the irony of a "frenetic standstill" (Rosa) in which the more quickly and extensively things change, the more they remain the same.

According to Rosa, the circle of acceleration works not only as a feedback loop, but also is driven by three external motors. The economic motor of capital accumulation drives the growth imperative of capitalist enterprise, with companies and industries maintaining a competitive advantage by accelerating technological innovation for the sake of increased productivity and higher profits. Under the logic of capital accumulation, a mutual escalation of quantitative growth and acceleration occurs: if more gets produced in less time through changes in technology and the timely reorganization (or elimination) of labor, then profit margins can be maintained and higher levels of consumption presumably can be enjoyed. Hence the process of "creative destruction" (Schumpeter) is subject continuously to speed-up pressures, since the spoils go to the swift and efficient. Faster flows of information, money, and goods enable the capitalist system to maintain a dynamic stability. "From a national-economic point of view," Rosa notes, "it

---

25. Ibid., 214.

is significant that the basic economic problem of a capitalist economy is not a (static) problem of distribution but rather one of maintaining accelerated circulation."[26]

A second, cultural motor—which Rosa dubs the "promise of acceleration"—complements the logic of capital accumulation. Historically, the modern impulse to accelerate arises after the gradual collapse of a Christian grand narrative and transcendent horizon, which in the medieval era had made secular time and worldly events seem unimportant compared to the question of eternal life and anticipation of Christ's second coming.[27] With the emergence of humanism, historical consciousness, liberalism, and the psychological subject in the modern era, the locus of meaning shifts to one's choices and experiences in *this* world, i.e., to self-determination and the quest for self-fulfillment in this life rather than preparation for the next. Rosa claims that moderns have come to embrace a secular version of hope for eternal life, one closely aligned with a secularized yet still resonant Protestant work ethic. Happiness is now associated with worldly success and a "full life," the promise of which can be realized through the efficient use of scarce time, i.e., by making smart choices within the limits of a single lifetime that has open to it an expanding array of life-experience options in a more mobile, materially prosperous society. For modern humans, fear of damnation is replaced by fear of being cheated by time and circumstances beyond one's control out of having as many "fulfilling experiences" as possible. This fear breeds a desire to master contingency through material wealth—and the security it allegedly affords—and also underlies the compulsion to compress time and multiply experiences so as to maximize the "richness" of one's psychological subjectivity. "Two techniques for supposedly increasing our experiential wealth have continually altered our everyday life in the last two centuries: multiplication and compression of the objects of experience . . . the more means of experience (TV programs, clothes, vacations, partners, etc.) we appropriate (*multiplication*), and the

---

26. Ibid., 164.

27. But see Noble, *Religion of Technology*. Noble shows how accelerated technological development has its historical origins in monastic practices and medieval theology, which valorized engagement in the mechanical arts as contributing to God's redemptive purpose for humanity, a finality conceived as the restoration of Adamic perfection. In one sense, Rosa's view of the current phase of hyperacceleration as unprecedented in history is unarguable, yet in light of the historical research done by Noble, Lynn White Jr., and others the contrast he draws between a static, medieval eternity and a dynamic, modern temporality seems facile. That said, Noble's observation that today's breathtaking pace of technological innovation is inspired in no small part by dreams of transcendence echoing back a millennium does not undermine Rosa's view of social acceleration as driven "structurally" by a feedback process and external motors.

more we concentrate them in time (*compression*), the richer our interior life will be—an increase in being through an increase in having."[28]

In reality, the dynamics of social acceleration render this hope vain, since the possible range of experiences for even a resourceful, unencumbered self continually shrinks in relation to the exponential growth of life-experience possibilities. As Rosa puts it, "contrary to the promise of acceleration, the *degree of exhaustion*, the ratio of worldly possibilities *realized* in a life to those that are *realizable*, continually *decreases* no matter how much we hurry around heightening the tempo of our lives."[29] In short, the promise of happiness through faster living is false; social acceleration turns the modern bourgeois quest for a "full life" into a Sisyphean task.

Nonetheless, the promise of self-fulfillment through fast living retains its appeal and is able to represent itself endlessly in myriad forms and images through the development of new advertising methods and marketing strategies. (A Google search for "happiness" currently yields around 365 million entries; on the all-important first page, we find that lucky item #7 titled "What is Happiness?—Finding True Happiness" is sponsored by Coca Cola.) Whether it is fast food, video games, hook-ups on Tinder, "flexible" volunteering, or packaged safaris, late capitalism excels at the delivery of canned experiences, the perceived value of which lies in no small part on how quickly and easily one can pop them open, consume the desired pleasure, and discard the empty container. Along with new kinds of synthetic drugs designed to induce pleasurable feelings, virtual and augmented reality technologies will extend the (false) promise of a psychologically "full life" still further in coming decades, enabling one to sample on demand a thousand and one lives—fictional and biographical, historical and futuristic—or easily and quickly alter one's perceptions and feeling-states in a thousand and one ways. *Westworld*, here we come!

Functional differentiation in modern society constitutes a third structural motor driving social acceleration because it magnifies the complexity of modern living. With demands on time, attention, and other resources coming from many angles, it becomes difficult for individuals and groups to

28. Gerhard Schulze, quoted in Rosa, *Social Acceleration*, 182. That capitalist culture falsely equates happiness with "having more" is hardly a new insight. From Wordsworth onward a host of artists, philosophers, and social critics have decried the alienating effects of commercialism. Erich Fromm's *To Have or to Be?* is a typical, twentieth-century example of this modern-lament genre, and Catholic social teaching since the 1960s has made the Marcelian theme of "having versus being" a staple of its critique of consumerism. The latest version appears in positive-psychology research by Martin Seligman and others on happiness, which documents the misery of those operating in the individualistic, hurry-up-to-have-more mode.

29. Rosa, *Social Acceleration*, 184.

synchronize processes, timetables, and events. The increasingly acute problem of *temporal desynchronization* takes many forms. As technology and economy race forward, for example, politics, law, and education come to be viewed as a worrisome "brake" on techno-economic progress. The time lag between what industry needs *now* and what legislatures, courts, and colleges can deliver in due time creates frustration for the former and puts pressure on the latter to "modernize" (i.e., streamline and speed up). Statutory and work-skill obsolescence thus becomes more common and more problematic. "In a high-speed society subject to constant change," William Scheuerman observes, "the half-life of even the most well-conceived legislative norms suffer from dramatic decline, as statutes become ever more outmoded at increasing rates."[30]

Individuals in search of the "full life" face time-related tensions and scheduling dilemmas that come with participation in multiple functional spheres. (Q: "How do I make time for family, work, soccer, church, neighborhood association, and political party—not to mention the latest on Netflix?" A: With another app!) Moreover, in late modernity growing numbers of people must cope with the blurring of previously well-defined times for different activities. Work becomes more flexible, project-oriented, and deadline-driven and thus spills over into nights and weekends, creating a more acute sense of time scarcity even as flexibility in fashioning one's schedule increases.[31]

The time squeeze also takes a toll on deliberative democracy as pressure mounts for social decisions to be made quickly and with only short-term consequences in view. Practices of citizenship and self-government require participation and sustained engagement, yet the time and other resources required are in short supply for large numbers of citizens. Rosa's disquieting thesis is that neither democratic movements nor counter-cultures can slow down, much less reverse, the self-propelling process of social acceleration, which now has gone into overdrive.[32] In late modernity the temporal

30. Scheuerman, "Citizenship and Speed," 298.

31. Wajcman, *Pressed for Time*. Wajcman argues that Rosa's "grand theoretical approach" misses too much sociology of everyday life; a more granular life-world approach discovers greater ambiguity in ordinary people's daily experience of technologies and the pace of life in contemporary society.

32. Rosa makes clear that not everything is continually going faster and faster. Along with various natural "speed limits" (geophysical, biological, and anthropological) and "islands of deceleration" (e.g., the Amish, though even their young are "deviced" now) within modern society, two forms of intentional deceleration—one functional, the other ideological—can be identified. The former consists of individual timeouts (e.g., a vacation, retreat, or ten-second mindfulness moment) from life in the fast lane for the sake of personal renewal, and thus functions as an *accelerator* by enhancing productive

gap between rapid technological and social change, on the one hand, and the comparatively slow pace of democratic decision-making and policy implementation, on the other, puts the latter at a structural disadvantage and encourages a reactive, situational politics devoid of long-range vision. Elected officials, administrators, and citizens face growing pressures to relinquish collective choice and regulation in favor of the faster, more flexible, and allegedly self-regulating economy. In other words, hyperacceleration favors neoliberal forms of governance and disfavors deliberative-democratic practices that aim to resynchronize natural and social ecologies by steering techno-economic progress in less risky directions and along pathways that respect geophysical, biological, and anthropological limits.[33]

Familiar fault lines of politics are shifting as technology races ahead, risk conflicts proliferate, and the pace of life quickens. Social philosopher Steve Fuller argues that the old left-right ideological divide is turning ninety degrees, as it were, and becoming a new up-down contest between techno-progressives and bio-conservatives. The former champion a proactionary agenda of exponential technology development and transhuman enhancement, while the latter look to the "wisdom of nature" and counsel a precautionary approach to technological innovation.[34] Where proactionaries see technological risks as opportunities to tinker and learn through trial and error, precautionaries view them as potential threats requiring a careful response focused on avoiding the worst possible outcome.

The parties involved in this debate represent rival epistemologies. Heirs to the Cartesian-Baconian tradition, techno-progressives acknowledge that nature is complex but deny that it is more complex than we can

---

capacity. The latter refers to countercultural (and often anti-modern) groups and movements (e.g., Luddites, slow food) that aim to slow down social-acceleration processes viewed as damaging to human values and authentic culture. However, in Rosa's view the historical record shows that ideological deceleration efforts don't succeed: "In the history of modernity, each wave of technological, organizational, or cultural acceleration at first encounters massive resistance and widespread skepticism but eventually triumphs, gradually silencing its critics" (300).

33. If Rosa is right about social acceleration, then is resynchronization even possible? In his judgment, our high-speed technological society is structurally incapable of slowing down and taming the growth imperative, and consequently massive ecological destruction (followed by social breakdown) is most likely our collective fate. Against this dark and deterministic outlook, I shall argue in Part II for a politics of engagement and time that, along with profound cultural renewal, might enable societies to operate within the limits of the planet's ecological boundaries while also sustaining material abundance and cultural dynamism.

34. Fuller and Lipinska, *Proactionary Imperative*, chap. 1. On the new bio-political landscape, see also the partisan yet still helpful discussion in Hughes, *Citizen Cyborg*, chaps. 6–10.

(eventually) know. By contrast, bio-conservatives believe we must learn to "live cautiously and precautionarily, cognisant of the fundamental limits to our contextual, perspectival knowledge and of the time-space indeterminacy of our actions' impacts."[35] While techno-progs are eager to speed things up, bio-cons want to slow them down. In world risk society, the proactionary-precautionary debate is moving toward center stage as scientists, entrepreneurs, policy-makers, and the public grapple with the effects, intended and unintended, of industrial and emerging technologies. I engage this debate in chapter 9, staking out a creative-middle approach capable of navigating between the Scylla of recklessness and Charybdis of relinquishment in technology development.

Three Meanings of Sustainable Development

In 2015, world leaders gathered under UN auspices to endorse an ambitious list of seventeen Sustainable Development Goals to be achieved by 2030. Yet beneath the surface of elite consensus, sustainable development remains a contested concept in theory and practice. In public-policy circles a range of social actors engage in discursive battles over what sustainability and justice require in various contexts. Differing priorities reflect incompatible background assumptions and disparate interests.[36] Without too much oversimplification we can identify three camps contesting the meaning of sustainable development: neoliberals, ecological modernizers, and green communitarians. Each camp offers a strategic vision and policy narrative for creating a more just, humane, and sustainable society, and each draws inspiration from a distinctive form of mythic, metaphorical imagination.[37] Here I analyze and compare the discourses employed by these camps in

---

35. Adam, *Timescapes of Modernity*, 35.

36. For a discourse analysis of sustainable development, see Dryzek, *Politics of the Earth*.

37. Harrison, *Constructing Sustainable Development*; according to Harrison, social actors employ policy narratives in order to define a social problem or set of related problems and offer solutions to it by drawing upon "a cohesive set of beliefs, meanings, understandings, technical concepts, and analytical methods" (4). Contending policy narratives provide material for discourse analysis, which in my view also must take into account the formative influence of a mythic, metaphorical imagination in shaping a sense of identity (personal and communal) and a normative vision of both past and future. Denying the power of mythic, metaphorical imagination or deconstructing its cultural expressions is bad hermeneutics; consequently, we are deaf to mythic narratives and blind to master metaphors that, for good or ill, provide direction and purpose in the movement of life. Put another way, narrative selves find meaning and co-create value and beauty within communities shaped by stories and metaphors.

light of world risk society, social acceleration, attitudes toward nature, and the proactionary-precautionary debate.

Sustainable Growth

The neoliberal camp, located politically within right-leaning parties and conservative think-tanks such as the Heritage Foundation, claim that a grand strategy of expanding markets across borders and limiting state intervention in the economy offers the best hope for "sustainable growth" worldwide. Competitive markets allocate resources more efficiently than socialist systems and generate the technological innovations that enhance productivity and raise living standards to the benefit of all. The apparent chaos of market activity is better viewed as a form of dynamic, spontaneous order. As a self-organizing and flexible system, the market allows individuals maximal freedom to pursue their interests (within legal-moral limits) and deploy their skills and know-how in the most productive way possible; it is a self-correcting system that eliminates inefficiencies in production through the discipline imposed by open competition. While a market-oriented approach typically results in widening inequality, neoliberals argue this side-effect must be tolerated for the sake of motivating people's best efforts and increasing society's aggregate wealth. To justify inequality, they invoke a familiar metaphor: a rising tide of market-driven prosperity lifts all boats. Egalitarian concerns about mal-distribution are misplaced, since growth in capital and market expansion serves the common good in the long run. Better to build a ladder of affluence all may climb in due time.

On the neoliberal view, every nation faces a clear choice—either increasing integration into the world market system or self-imposed isolation—and hence a clear prospect—either growing prosperity or endemic poverty and stagnation. The favored policy mix includes trade liberalization, privatization, deregulation, balanced budgets, strong protections for property rights, and a shrinking of the welfare state. Neoliberals strongly support the WTO and multi-lateral efforts to further open markets across the globe, e.g., through NAFTA and, more recently, the Trans-Pacific Partnership. Endorsing some version of Rostow's modernization model, neoliberals prescribe growth-oriented strategies for the techno-economic development of poor countries that continue to "lag behind." North-South inequalities are said to be a function of limiting factors—cultural and political—within less-developed nations, not the historical result of patterns of forced dependency and domination.

At whatever scale one might consider, neoliberals hold that environmental problems are best handled by market mechanisms. They point to studies that lend credence to the "intensity of environmental wear hypothesis": a developing country in the early stages of growth will generate higher levels of pollution—up to a point; further economic growth will then result in decreasing levels of environmental damage over time as a now wealthier nation gains the wherewithal to go green by imposing environmental regulations and adopting cleaner technologies. In short, increased wealth is the key to environmental health.[38]

Many neoliberals agree with Gifford Pinchot (architect of the "maximum sustained yield" policy) who once declared that only two things—people and natural resources—exist on this earth. Under the tenets of free-market environmentalism, if natural resources are privatized and price signals made clear, then any scarcities that arise will trigger a timely and effective search for either a more efficient use of the (limited) resource in question or a suitable substitute. Underlying this market logic is a *Promethean-cornucopian imagination* that revels in individual freedom, human ingenuity, and dreams of unparalleled material prosperity.

Neoliberals tend to reflect the attitudes characteristic of what Beck calls early modernity, which accepts with few reservations the twin dogmas of material progress and domination of nature. Accordingly, the limits-to-growth thesis (Ehrlich, Meadows, et al.) is rejected in favor of a Promethean vision in which continual population increase and economic expansion complement one another. Human intelligence, Julian Simon argues, is the "ultimate resource" capable of overcoming any environmental challenge or biophysical limit.[39]

Neoliberal demands for deregulation, a more flexible workforce, and a more "relevant" educational system attuned to business needs are motivated by a desire to optimize and speed up production and thus increase profits, which in turn can finance more production and raise living standards in a "virtuous spiral" of sustainable growth. Environmental regulations become necessary only when a clear, scientifically confirmed danger to human health is present; when scientific uncertainty exists, they tend to take a wait-and-see approach on the assumption that greater harm comes from governmental constraints on economic activity. This outlook aligns many (but not all) neoliberals with techno-progressives, who take a proactionary

---

38. Finn, *Just Trading*, 163–67. As Finn notes, neoliberals often downplay the necessary role of government in establishing environmental standards. I would add that, in many cases, state environmental policy does not reduce or eliminate aggregate levels of pollution, but rather creates a market that results in its relocation/displacement to locales deemed inexpensive and pliant politically, i.e., to remote rural areas and low-income countries.

39. Simon, *Ultimate Resource 2*.

approach to technology development and favor the freedom to innovate and experiment with promising new technologies, even risky ones, provided the risk-benefit analysis is backed by "sound science."[40]

To sum up, for neoliberals the discourse of sustainable development is best translated into an optimistic storyline of "sustainable growth" through the unleashing of productive potential and entrepreneurial spirit. We can solve environmental problems and enjoy unmatched prosperity, they believe, by getting the bureaucrats out of the way and encouraging individuals to take ownership of their own destinies.

Pollution Prevention Pays

Ecological modernizers are situated politically among left-leaning forces in EU countries, while also occupying a seat at the policy-making table in China, Japan, Brazil, and South Africa.[41] They seek a marriage of sustainability and increasingly de-materialized economic growth made possible by strong environmental standards, integration of economic and environmental policies at national and global levels, and a swift transition to clean technologies and closed-loop production models in industry, agriculture, transportation, and other sectors.[42] Here the operative meaning of sustainable development is specified by the concept of ecological modernization, the core dynamic of which is a market transformation driven by a synergy of government policies (e.g., carbon taxes, feed-in tariffs to stimulate renewable energy), progressive business behavior ("corporate sustainability" and socially responsible investing), and more consciously green consumer behavior. Supporters argue that an eco-modernizing strategy not only would "open up new markets and create new demands; if executed well, it would stimulate innovation in methods of production and transport, industrial organization, consumer goods, in short, all those elements that Schumpeter once identified as the forces that produce the 'fundamental impulse that sets and keeps the capitalist engine in motion.'"[43]

---

40. A caveat: some economic neoliberals hold socially conservative views, in which case they may side with bio-conservatives on certain technology issues, e.g., stem cell research. The new up-down politics has the effect of remixing or shaking up the older left-right categories and creating odd bedfellows.

41. For China, see the *China Modernization Report 2007—A study on ecological modernization*. Accessed January 23, 2016, http://www.modernization.com.cn/cmr2007%20overview.htm.

42. On eco-modern design thinking, see Hawken et al., *Natural Capitalism*; also McDonough and Braungart, *Upcycle*.

43. Hajer, *Politics of Environmental Discourse*, 31–32.

The eco-modernizer's mantra—pollution prevention pays—is attractive politically because it presents policy-makers and public alike with a positive-sum game to play, rather than being saddled with the apocalyptic end-game scenarios of radical environmentalists. While the specter of overshoot and collapse is taken seriously, ecological modernizers see no need for collective sacrifice and austerity. The dilemmas posed by impending biophysical limits can be dissolved by a set of win-win policy solutions and "planetary management." Beneath this logic is the *enlightened technocratic imagination* of elites that seek to guide "spaceship earth" (Fuller) through its current turbulence with good governance and clean technology.

With the neoliberals, ecological modernizers welcome the opportunities opened up by economic globalization; both accept democratic capitalism and technologically driven social acceleration as a given. At the same time, advocates of ecological modernization reject neoliberal over-reliance on market solutions and insist that government at several levels must play a leading role in greening economy and society. A structural transformation of the economy is both necessary for environmental reasons and desirable from a progressive business perspective attentive to the profitability of green products and services.[44] Private-sector examples include Al Gore's Generation Investment Management, Atlanta-based flooring manufacturer Interface, and the consumer-product company Seventh Generation—all of which aim to show that "sustainable capitalism" is not oxymoronic by outperforming the market with a long-term strategy and stringent sustainability criteria.[45] Ecological modernization theory "recognizes the structural character of the environmental problematique but none the less assumes that existing political, economic, and social institutions can internalize the care for the environment."[46] On this view, markets can serve societies best within a robust framework of governance that establishes strong regulations to protect workers, consumers, and natural ecologies yet also encourages the flexibility and innovation necessary for making a market-oriented transition to environmentally sustainable production and consumption.

Within the EU, ecological-modernization theory and policy-making has moved in a progressive direction since its inception in the early 1980s. Early eco-modern thinking generally limited the scope of its analysis to advanced-industrial nations and emphasized technological solutions in

44. For an example of an eco-modern policy agenda at the national level, see the Green New Deal Group, "A National Plan for the UK."

45. On Generation Investment Management's sustainable business strategy, see Fallows, "Secrets of Al Gore." For an insider's account of Interface's "sustainability journey," see Anderson, *Radical Industrialist*.

46. Hajer, *Politics of Environmental Discourse*, 25.

dealing with environmental problems. Today's more progressive ecological modernizers promote open, participatory decision-making at all levels of governance (e.g., citizens' conferences in Denmark); focus more on the possibilities for a broader transformation of practices (e.g., LEED building design, full-cost accounting), legal frameworks (e.g., Paris Agreement), and institutions (e.g., environmental education) to ensure long-term sustainability; and recognize the need to build out a smart, post-carbon infrastructure for the twenty-first century.[47]

The progressive eco-modern agenda is grounded morally in notions of environmental rights, including the rights of future generations: "Intergenerational solidarity in dealing with the sustenance base has emerged as an undisputed core principle."[48] In situations of risk and uncertainty, where the impact of current activities on future generations could be catastrophic, ecological modernizers usually follow the maximin rule—a decision rule for avoiding worst-case scenarios—and thus invoke the precautionary principle. Against neoliberals, they argue that policies and practices guided by a precautionary approach will yield both economic gains and societal well-being in the long run. In short, what is prudent can also be profitable. While neoliberals tend to make early-modern assumptions about technological progress and the calculability of risks, eco-modernizers generally adopt a more critical, late-modern stance toward claims of prediction and control. At the same time, the two camps share modern assumptions regarding science-led progress and economic rationality.

In sum, the hopeful vision articulated by ecological modernizers turns on the belief "that the capitalist political economy needs conscious re-configuring and far-sighted action so that economic development and environmental protection can proceed hand-in-hand and reinforce one another."[49] Informed by science and armed with new technologies, responsible elites in business and government can cooperate and lead society through the necessary transition to sustainability.

Just Sustainability

A more diverse grouping politically, green communitarians include among their number radical environmentalists, ecological economists, autonomist Marxists, progressive NGO and labor leaders, environmental-justice activists, bio-regionalists, eco-feminists, organic and traditional farmers,

---

47. See, e.g., Rifkin, *Third Industrial Revolution*.
48. Mol and Sonnenfeld, "Ecological Modernisation," 7.
49. Dryzek, *Politics of the Earth*, 143.

creatives, and representatives of peasant groups and indigenous peoples. Within government, greens have gained footholds in countries with proportional representation and often aligned themselves with progressive eco-modernizers. However, many deep-green communitarians gravitate toward movement politics or focus on alternative institution building within civil society. As mentioned above, these groups and movements remain in dialogue with each other and occasionally lock arms in direct action against neoliberal policies that, in their view, critically endanger the life systems of the planet.

Inspired by the "small is beautiful" approach of Gandhi, E. F. Schumacher, and others, green communitarians espouse egalitarian visions of "just sustainability" based on a decentralized, pluralist approach to governance and a local/bio-regional model of cooperative economy. They embrace an ethos of local self-reliance, communal sharing, and social innovation while advocating for steady-state economic practices and participatory democracy. The values of sufficiency, simplicity, earth-care, solidarity, and service to others reflect a commitment to community and a more holistic, spiritually oriented view of human flourishing. Arguing for the (relative) virtues of self-sufficiency and human-scale institutions, greens reject all forms of concentrated power, whether it is the corporate-dominated agenda of neoliberals or the ecological modernizers' elite, planetary-management approach.

Green communitarians agree with ecological modernizers that the mega-hazards of world risk society require strong precautionary measures. Both camps give more weight to intergenerational justice and the "rights of nature" than do neoliberals, whose circle of moral concern typically does not extend beyond humanity here and now. Still, important differences remain in their respective attitudes toward nature and stances within the proactionary-precautionary debate.

Persuaded that humanity has entered the Anthropocene, many eco-modernizers accept the view that we now inhabit a post-natural world, i.e., an "eaarth" (McKibben) profoundly and irretrievably altered by the modern trifecta of technological innovation, economic growth, and population explosion. The Great Acceleration since 1950 is an unalterable fact of contemporary life, and so prudence dictates active management of technological and social change with the aim of reducing humanity's ecological footprint. On this view, nature has become a complex set of hybridized environments, a shifting and sometimes volatile, dangerous mix of the natural and the artificial. Consider, they say, how only some toxic wastes and organic pollutants get cordoned off or incinerated; a good deal of them—in particle and other forms—are released in myriad daily activities of industrial society and remain in the land, air, and waters—and in living bodies—with health effects that epidemiologists are just beginning to comprehend. If the planet effectively is "man-made" from now on, then conscious choices about constructions of "nature" can and should be confronted head-on. As biologist David Lodge and historian Christopher Hamlin put it, "it is not a matter then of doing things nature's way, but rather of deciding which of nature's ways or forms we want to establish, maintain, restore or change."[50] This implies, further, that older ideas of preserving pristine Nature no longer make sense. From now on we have to monitor and manage a complex, ever-changing hybrid world—a daunting but unavoidable challenge that requires a new and more integrated model bringing together science, ethics, and public policy (most recently under the banner of sustainability science). What follows practically for ecological modernizers is a judicious application of the precautionary principle coupled with a tempered, techno-progressive approach to managing the planet.

By contrast, many green communitarians adopt a bio-conservative stance and thus reject or heavily qualify the end-of-nature thesis. Often influenced by or deeply sympathetic to indigenous worldviews and traditional wisdom, green communitarians protest against violations of sacred landscapes and seek to protect pre-modern lifeways. Unlike neoliberals and eco-modernizers, they refuse to fuel the external motors of social acceleration. The post-capitalist society they envision will not be caught up in the circle of acceleration; rather, it will resynchronize natural and social ecologies through regenerative practices.[51] Animating this countercultural logic is an *arcadian imagination* that draws on various sources. Some greens adopt a deep-ecological attitude toward nature—a more-than-human world to which persons owe respect, even reverence. Others take inspiration from the new sciences of chaos, complexity, and connection, which suggest the need for a paradigm shift from an objectifying, mechanistic mindset to an ecological worldview emphasizing the interrelatedness of all things.[52] Many believe it is both possible and highly desirable to cultivate mutual, dialogical relationships with nature-as-subject. Wildness not only still exists, it holds within itself the "preservation of the world" (Thoreau).

While some overlap exists between the policy proposals of eco-modernizers and green communitarians—e.g., both call for adoption of closed-loop production processes and other forms of bio-mimicry in designing industrial systems—the latter offer more radical critiques of and alternatives

---

50. Lodge and Hamlin, quoted in Jenkins, *Future of Ethics*, 155; see also Biello, *Unnatural World*.

51. Wahl, *Regenerative Cultures*.

52. Prigogine and Stenger, *Order Out of Chaos*; Kaufmann, *Self-Organization and Complexity*.

to liberal capitalism that draw on a variety of perspectives at the margins. Indian eco-activist Vandana Shiva, for instance, denounces the inherent violence of the current system and deems it incapable of reform from within: "The global corporate economy based on the idea of limitless growth has become a permanent war economy against the planet and people . . . The present global 'war' is the inevitable next step for economic and corporate globalization driven by a handful of corporations and powerful countries that seek to control the earth's resources and to transform the planet into a supermarket in which everything is for sale."[53]

Socially located among the poor and vulnerable, green communitarians are attentive to economic globalization's dark side. They are partial to radical social analyses such as Immanuel Wallerstein's world systems theory, which holds that the current international division of labor creates a situation of systemic injustice on a vast scale.[54] According to Wallerstein, the growth imperative of historical capitalism—i.e., its relentless drive to accumulate capital and expand markets—is the systemic root not only of global injustice but of mounting environmental crises as well.[55] The Great Acceleration has thrown into sharp relief a "second contradiction" of the capitalist world-economy, namely its tendency to over-exploit nature and thus endanger the conditions of production upon which it depends. In response to such long-term trends as de-ruralization and rising urban labor costs—the latter only partially offset by technological innovation (and, more recently, re-ruralization of some industries)—capitalists increasingly find that they can maintain desirable rates of profit most easily by externalizing environmental costs. In the context of democratization, the ensuing despoliation of nature provokes popular demands for environmental protection, but states quickly run into structural constraints in their attempts to balance economic interests and environmental values. For one, capitalists fiercely resist environmental regulations that cut too deeply into profits, and presently mobile capital can in many instances outflank regulatory powers. For another, governments that fail to create a "friendly business environment"—i.e., one with low social and environmental costs—will be voted out of power with the next economic downturn, for capitalist society cannot envision any collective option other than continuous growth. Consequently, an ever-expanding capitalist world-economy continues to generate pervasive externalities that, cumulatively, threaten to degrade or destroy planetary life systems over time, despite efforts at piecemeal reform

53. Shiva, *Making Peace*, 3–4.
54. Wallerstein, *Capitalist World-Economy*.
55. On the systemic character of the growth imperative and "treadmill of production" in capitalist society, see also Gould et al., *Treadmill of Production*.

and rising public concern. In short, whether predicated upon neoliberalism or ecological modernization, "sustainable capitalism" is a contradiction in terms, with no exit to the growing eco-social crisis available within the confines of the present world system.[56]

Given their historical and structural analysis of the global eco-social crisis, many greens conclude that a commitment to radical environmental politics must take priority over a cultural-transformation strategy aimed at promoting a shift to green consciousness and more eco-friendly lifestyles. They recognize that "eminently rational appeals on the part of 'us' to change our attitudes, or lifestyles, so as to advance a general 'human interest' are liable to be ineffective. This is not because (or *primarily* because) 'we' are irrational, but because the *power* to make a significant difference, one way or the other, to global, or even to local environmental change is immensely unevenly distributed."[57] Adopting new values and lifestyles is necessary—indeed, it is crucial—but not sufficient for making the transition to a more just, humane, and sustainable society. Thus for green communitarians the storyline for our time is the ethical-political struggle for "just sustainability," a long-term struggle involving both resistance to neoliberal globalization and a Gandhian constructive program of practical experiments in cooperative economics and participatory democracy at the local/bio-regional level. This strategy aims to *displace* rather than reform a dysfunctional world system through alternative institution building while also organizing to resist and *disrupt* the current system's worst abuses through nonviolent action.

---

56. Wallerstein, *End of the World*; eco-modernizers argue that Wallerstein ignores the possibility of a market transformation to high levels of eco-efficiency and a dematerialized economy. Wallerstein counters that reforms within the current system only postpone the inevitable, because the "systemic crisis has narrowed in various ways the possibilities of capital accumulation, leaving as the one major crutch available the externalization of costs. Hence, it is less likely today than ever before in the history of this system to obtain the serious consent of entrepreneurial strata to measures fighting environmental degradation" (85). Yet as Beck notes, in world risk society the "entrepreneurial strata" are divided: some support eco-modern reform, others resist it, depending on their material interests. From another angle, Herman Daly and other ecological economists agree with Wallerstein's conclusions: the growth imperative of global capitalism cannot be reconciled with the inherent limits of the biosphere's carrying capacity, and so a paradigm shift to some type of steady-state economy is necessary. For a similar analysis within the UK context, see Jackson, *Prosperity without Growth*.

57. Redclift and Benton, "Introduction," 7.

Which Strategic Vision and Policy Narrative?

What are we to make of the ongoing sustainable development debate? While the neoliberal agenda of "sustainable growth" continues to hold sway in many halls of power, it has been challenged with some success by eco-modernizers, most notably with the recent passage of the Paris Agreement on climate change and Kigali accord to phase out HFCs (a highly potent greenhouse gas). The parameters of mainstream political discourse and policy debate continue to be shaped by these two camps. However, given the severity and systemic nature of the global eco-social crisis, I suggest the more important dialogue is between progressive eco-modernizers and green communitarians committed to alternative practices and institutions. The basic question, then, is this: Which strategic vision and policy narrative?

Some historical perspective may prove helpful in thinking about the options before us. From 1750 onward the British countryside started to empty out as farming increasingly became a mechanized affair and both land and labor were transformed into commodities within a growing industrial economy. In *The Great Transformation*, historian Karl Polanyi interprets the wrenching transition from field to factory as a "double movement." First, the allegedly free, self-regulating market sphere envisioned by (classical) liberal thinkers and constructed by industrialists generated massive social dislocation. Second, in response to urbanization and its ills, society sought to protect itself through mutual aid, union formation, social legislation, and other means.[58] On the political battlefield, the powerfully disruptive forces of economic liberalism were met by socialists and revolutionary movements.

Perhaps we are in the midst of another Polanyian double movement leading to changes in practices and policies that provide just enough environmental protection and social stability to allow capitalism in some altered form a continuance of its 500-year run—at least for the foreseeable future. While successful systemic adaptation may occur, the historical and structural analyses put forward by Wallerstein, John Bellamy Foster, James O'Connor, Jason Moore, Paul Mason, and others suggest the waters and weather ahead will be much darker and more turbulent than, say, the World Economic Forum crowd can imagine.[59]

Considering the basic question—which strategic vision and policy narrative?—it appears we have three options once the moral and practical failures of the decades-old neoliberal project are recognized. One option

58. Polanyi, *Great Transformation*.

59. Foster, *Vulnerable Planet*; O'Connor, *Natural Causes*; Moore, *Web of Life*; Mason, *Postcapitalism*.

demands a clear choice and full commitment to either ecological modernization or green communitarianism. Thus, eco-modernizers such as Al Gore, Bill Gates, Elon Musk, Ted Nordhaus, and Michael Shellenberger stump for more green R&D funding to accelerate the envisioned transition to a highly-efficient post-carbon economy, while green communitarians press for separation from a capitalist system bent on using its considerable powers to co-opt or domesticate opponents. For the latter, it is better to avoid the inevitable compromises that come with participation in representative government, corporate-NGO partnerships, and the like and focus instead on building alternative institutions from the ground up.

A second option adopts a both-and approach to systemic change based on dialogue and cooperation between progressive eco-modernizers working inside the system and greens working outside established power centers. This inside-outside strategy presumes a renewal of democratic energies, works on many fronts, and envisions a transition to a diverse, hybrid economy and post-jobs society geared to sustainable abundance for all.

The third option aligns closely with the second but points up the intransigence of existing power structures and limits of dialogue, collaboration, and social-learning approaches to systemic change under political conditions marred by corruption, plutocracy, and authoritarian control. Hence the need at critical sites for practices of nonviolent resistance to projects and policies and that harm the poor and despoil the earth. Business as usual must be disrupted, suffering must be made visible, and a creative tension must be generated for radical reforms to take hold. Strategically, this option adopts a transformational approach to systemic change involving three elements: gradual *displacement* of neoliberal hegemony through alternative-institution building; *radical reform* at all levels of governance aimed at a wider distribution of economic resources and political power; and nonviolent *disruption* of the current system's most exploitative, harmful practices through creative forms of direct action. To be clear, a three-pronged transformational strategy does not aim for total replacement of one system by another, a prospect neither possible nor desirable in light of the bloodiest century just passed. Rather, it envisions a shift over time from the current configuration of power relations, social practices, and institutional models to a more diverse, hybrid economy and socially inclusive post-jobs society that promises to be, among other things, more resilient in the face of increasingly difficult environmental conditions precisely because it has found ways to distribute power and resources more widely and equitably.

Over the next three chapters, I critically examine the church's evolving response to industrial modernity and an emerging world risk society, and specifically its evaluation of competing sustainable-development discourses and initial response to the emerging proactionary-precautionary debate. I show how Catholic social teaching has favored a both-and strategy toward sustainable development (option #2), affirming the perspectives and concerns of both eco-modernizers and greens and thereby encouraging dialogue and cooperation between them. In Part II, I suggest that socially conscious Christian leaders do well to pursue a three-pronged transformational strategy (option #3), whether such leadership is exercised in church-related social-action projects (e.g., New Jersey-based GreenFaith) or secular organizations and movements.

3

The Church's Ambivalent Response to Techno-Economic Progress (1891–1981)

Catholic social teaching on integral human development and the nexus of technology, economics, politics, and ecology in a globalizing world has evolved over many decades. While this chapter and the next two focus mainly on ecclesial responses to the rise of world risk society from 1950 to the present, the background story is the church's ambivalent response to modernity dating back to the Enlightenment, the "shock" of the French Revolution, and the destabilizing impact of early capitalism on a rural Catholic aristocratic tradition in England and the Continent.

Modern Catholic social teaching has developed in response to creative currents of non-official Catholic social thought and action—or simply, social Catholicism—as well as societal trends, world events, and streams of thought in the wider culture. A teaching church often critiques what it sees as ideological distortions and dangerous tendencies in modern society, e.g., the failures of liberalism and socialism, or the "culture of death" that sanctions abortion, euthanasia, capital punishment, and excessive military spending. At the same time, a learning church has been enriched by social Catholicism's prophetic voices and engagements with modern social movements—in particular struggles for economic justice, peace, democracy, and human rights—as well as insights from modern science and philosophy (e.g., Buber's philosophy of dialogue). From the 1960s onward, church leaders have engaged in a sometimes fruitful, sometimes frustrating dialogue with feminist, ecology, and anti-racist movements within church and society.

Keeping in view the discussion of a risky, runaway world and sustainable development's contested meanings, I explore several questions over the next three chapters. Most broadly, how has Catholic social teaching responded to techno-economic progress and its dark side in the modern era? More specifically, what social theories have been presumed or deployed as church leaders read the signs of the times and offer moral norms, social principles, and broad policy recommendations aimed at fostering authentic human development? What background theories, paradigms, and forms of imagination have inspired and informed an evolving Catholic social-ethical vision? In recent decades, how has both papal thought and episcopal teaching at the national/regional level engaged the sustainable development debate? Which policy narratives have been favored or frowned upon? And finally, what stance has the church adopted in relation to the emerging debate between proactionaries and precautionaries over industrial and emerging technologies? Addressing this nest of questions will enable us to raise both substantive and methodological issues bearing upon the sustainable development, as it were, of Catholic social thought and action in coming decades.

This chapter looks first at the origins of modern Catholic social teaching with attention paid to the vital role of nineteenth-century social Catholicism in shaping a papal social-encyclical tradition that in turn inspired new forms of social Catholicism, notably the British distributists, during the early-twentieth century. I then review ecclesial responses to rapid technological development and an emerging world risk society from mid-century to the early 1980s, noting key emphases as well as three tensions in the ecclesial documents of this period. I also discuss church teaching on work and leisure as well as liberation theology, given the importance of these topics for arguments advanced in Part II.

Social Catholicism and Leo XIII's Strategic Response to Industrial Capitalism

Pope Leo XIII's response to the social question—i.e., the scandal of grinding poverty amidst gross inequality experienced by a newly industrialized labor force—came after decades of social dislocation and political turmoil in nineteenth-century Europe. The double movement of liberal-capitalist initiative and communal-protectionist response analyzed by Polanyi was the context for the emergence of social Catholicism in Germany, France, and elsewhere. A creative minority within a conservative church, social Catholics established Christian factories, mutual-aid societies, credit unions, youth organizations, and worker associations in response to harsh working conditions and widespread suffering. In study circles, most notably the Fribourg Union, activated clergy and laypersons debated a range of alternatives to liberal capitalism, drawing on Aquinas to formulate an ethical-practical and philosophical framework for addressing the social question. Leo's encyclical *Rerum novarum* (1891), the charter document of modern Catholic social teaching, was informed by social Catholicism's various activities and incorporated major elements of social Catholic thought.[1]

In *Rerum*, Leo called for a new concord between owners and laborers structured around reciprocal rights and duties. The pope championed the rights of industrial workers to receive a living wage, form unions, and strike when necessary; such rights conform to natural law and the *telos* of persons flourishing within a just, peaceful community (RN #11–17). Leo held out the hope that a just wage would enable frugal factory workers not only to sustain families but also to save and eventually own property (#65–66). Social peace could be secured through a wider and more just distribution of earth's abundance, conceived here as productive property that is privately owned yet also subject to norms of just use for the sake of the common good. Grounded theologically in the universal destination of created goods (i.e., God's good creation is given for proper use to all humanity), the desirability of a wide distribution of wealth and economic opportunity throughout society was reaffirmed and extended in later papal social teaching to include calls for workplace democracy and worker-owned cooperatives.

In Leo's teaching the state, while responsible for protection of the common good, must respect individual rights and freedoms grounded in natural justice. A properly functioning state plays a subsidiary role in helping local communities and civil-society groups meet the needs of its members, intervening only when necessary to remedy a social evil, e.g., unsafe working conditions (RN #36). Forty years later, amidst the Depression and socialist calls for state control of the economy, Pius XI reaffirmed in *Quadragesimo anno* the principle of subsidiarity: "It is a fundamental principle of social philosophy, fixed and unchangeable, that one should not withdraw from individuals and commit to the community what they can accomplish by their own enterprise and industry" (QA #79).

The principle of subsidiarity serves as a good example of how papal social teaching first was informed by and in turn provided ecclesial support for various experiments in reform and community provision of goods undertaken by social Catholics. Commenting on the historical context for articulating the idea of subsidiarity, Vincent Miller notes:

---

1. Misner, *Social Catholicism*.

> The concept arose out of the profound, concrete experience that revolution had overturned the social order. As capitalism "freed" land to be put to more productive use, it dissolved relationships between gentry and peasantry and created a new social reality of class. The peasantry was turned into a working class, always vulnerable to unemployment and insufficient wages. Social Catholics could not simply restore what had been lost. They had to create new forms of community, new structures of support and action, in a radically changed context.[2]

While subsidiarity and other social principles enunciated by Leo and Pius were drawn primarily from Aquinas, the *carrier* of papal teaching in society was a small but influential social Catholic movement responsive to the vulnerabilities and degradations experienced by the new working class. From the various practical experiments in social reform and mutual aid, a distinctively Catholic form of social thought and action emerged.

Leo XIII rejected class struggle and focused much of his energy in *Rerum* on a rapprochement between capital and labor. According to Joe Holland, Leo's reformist strategy was an attempt to reinforce loyalty to the church among Europe's Catholic working class (particularly men) while forming an alliance with moderate liberals against socialist forces. The pope initiated a strategy of accommodation to modern democracy that paved the way for (or at least acquiesced to) active Catholic participation in an emerging Christian-Democratic political movement. In addition, Leo's strong endorsement of the Thomist revival underway in ecclesial circles was intended to bolster support for the papal critique of secular-modernist assumptions spawned during the Enlightenment; it also would provide the philosophical basis for a Catholic affirmation of modern democratic ideals, as evidenced in the writings of Jacques Maritain and other twentieth-century Christian humanist thinkers. On Holland's reading, the three-pronged Leonine strategy of social Catholicism, Christian Democracy, and Thomism remained in place, albeit with some important modifications, until around 1960.[3]

A *communitarian imagination* underpinned both social Catholicism and papal social teaching's vision of a restored Christian society. As Miller points out, the Catholic communitarian ethos and imagination—and the organic social vision it expressed—is rooted in medieval nostalgia:

> [T]he tradition rests upon a very romantic, and not entirely accurate, imagination of the medieval world. This is the vision of the organic social order, where persons are not individuals, but

---

2. Miller, "Catholic Social Teaching."
3. Holland, *Modern Catholic Social Teaching*, chap. 4.

remain in a network of sustaining relationships of intertwining rights and responsibilities. The bounds of fealty between peasants and nobles, craft guilds, and so forth embodied these ideals only imperfectly. This romantic imagination of a past that never quite was forms the metaphorical basis for the communal conception of the human person and our notions of the common good.[4]

A deep current of romantic conservatism flows through Catholic culture, providing the metaphorical-affective basis for affirming the social nature of humanity, a human nature expressed primarily through gift exchange, cooperation, and conviviality rather than through competitive striving for wealth and status. Its provenance in medieval nostalgia aside, Miller's helpful observation highlights the importance of an "ethical imagination"[5] in forming our conceptions of a worthy life and a good society. Steeped in a communitarian ethos and imagination, a distinctively Catholic social-ethical vision of human flourishing has challenged liberal-capitalist and state-socialist assumptions alike.

The British Distributists

In the early-twentieth century, Eric Gill, G. K. Chesterton, Hilaire Belloc, Fr. Vincent McNabb, and other British distributists sought to promote and practice papal social teaching. At once reactionary and radical, Catholic distributists held that neither liberal capitalism nor state socialism could create a good society, because both ideologies flouted the demands of distributive justice, ignored the principle of subsidiarity, and failed to respect human dignity and the transcendent dimension of existence. Founded in error upon materialist metaphysics (dating back to the atomism of Epicurus), in practice these systems could not help but bring misery and enslavement. Addressing the situation in England, distributist voices were especially critical of industrial capitalism. Stripped of land and tools by the enclosures and factory system, people were forced into wage slavery, or what Belloc referred to as the "Servile State," a grossly unjust system in which a few owners enjoyed real freedom while the majority remained trapped in insecure, soul-deadening jobs.

In counterpoint to the enslaved factory worker, Gill held up the figure of an independent artisan dedicated to creating high-quality goods that are as beautiful as they are useful: "The type of the free workman is the artist

---

4. Miller, *"Catholic Social Teaching."*
5. Kearney, *Wake of Imagination*, 361–66.

... All free workmen are artists." Commenting on Gill's ideal, John Hughes notes, "The artist is the paradigm of the free worker because he has freedom in expressing himself and is responsible for what he creates, even if commissioned by another; he also takes delight in his work, and this delight is manifest as beauty in the thing made."[6] For Gill, David Jones, and other Catholic artists, distributism offered an alternative to both the utilitarian ethos of modern industry and Romantic expressive individualism. With Hilary Pepler, Gill formed the Craft Guild of St. Joseph and St. Dominic (1920–1989); guild members understood the main purpose of creative manual work was growth in the love of God, and the utility and beauty of well-made things were intended always to reflect the glory and goodness of God rather than the creative genius of a "heroic" artist.

For British Catholic distributists, control over the means of production was a precondition for the exercise of responsible freedom and artisanal creativity. Hence they called for the widest possible ownership of productive property—in the form of family farms and local cooperatives—as the surest safeguard of political and economic freedoms. Importantly, production ought to be for *use*, not profit; what things are good for only makes sense within the context of community life and divine worship. As an integral aspect of the human vocation to love and serve God and one's neighbors, economic life is subject to moral norms and does not constitute an autonomous sphere within society.

Essentially a back-to-the-land movement, the distributists held that rural communities and mid-sized towns are the optimal form of civilization. Nostalgia for the presumed organic harmony of medieval society permeated their thought and aesthetic sensibility. The ideal was Benedictine stability, the discipline and rhythm of *ora et labora*, and a reuniting of the useful and beautiful. In sum, they envisioned a cooperative, communitarian society in which, as Gill put it, "the artist is not a special kind of man but every man is a special kind of artist." While the distributists were small in number and remained at the margins of English Catholic life, they influenced later social-Catholic thought and action, e.g., the U.S. Catholic Workers, Mondragon's worker-owned cooperatives, and the green-communitarian writings of E. F. Schumacher.

6. Hughes, *End of Work*, 183.

The Pope of Technology

Distributist ideals also found a place in the social teaching of Pius XII, who expressed support for cooperative enterprise and reforms aimed at workplace democracy. In a 1944 broadcast message, Pope Pius stated that

> [t]he small and average sized undertakings in agriculture, in the arts and crafts, in commerce and industry, should be safeguarded and fostered. Moreover, they should join together in co-operative associations to gain for themselves the benefits and advantages that usually can be gained only from large organizations. In the large concerns themselves there should be the possibility of moderating the contract of work by one of partnership.[7]

This papal "directive" was reaffirmed by John XXIII in his social encyclical *Mater et Magistra* (#84), and similar statements supportive of the democratization of wealth and economic opportunity are to be found in every major social encyclical up to *Laudato*.

Led by Christian Democratic parties whose platform was informed by Catholic social teaching, the postwar reconstruction of European economies moved forward rapidly. In this context, the church's cautiously optimistic position on modern technology found its first, full articulation in the many public addresses of Pius to professional groups in various scientific and technical fields. The pope strongly encouraged progress in science and technology even as he warned against a utilitarian ethos and "technological spirit" (i.e., technocracy) coming to dominate society. For Pius (and later popes), the nuclear giant–ethical infant dilemma articulated by Einstein, Bradley, and others was a recurring theme.[8] The work of redemption in modern society entailed closing the gap between growing technological power and humankind's moral frailties; a restoration of Christian values would resolve the contradiction between science-led material progress and the threat of nuclear self-destruction.

---

7. Pope Pius XII, broadcast message on September 3, 1944.

8. Among many statements, perhaps the best-known is that of Gen. Omar Bradley: "We have men of science, too few men of God. We have grasped the mystery of the atom and rejected the Sermon on the Mount. The world has achieved brilliance without conscience. Ours is a world of nuclear giants and ethical infants. We know more about war than we know about peace, more about killing than we know about living. If we continue to develop our technology without wisdom or prudence, our servant may prove to be our executioner." General Omar Bradley, "Armistice Day Speech (November 11, 1948)."

While potentially enslaving, Pius held that modern technologies opened up "limitless possibilities." Hence Christian technologists had a responsibility to apply themselves diligently: "The call to be a Christian is not God's invitation to enjoy the aesthetic pleasure of contemplating His wonderful order; it involves a tireless and self-sacrificing effort in every sphere . . . of life."[9] The pope affirmed the vocation of scientists and engineers to build up the earthly city as an anticipation of the City of God, often reminding his listeners of their responsibility to use their talents and work output for the sake of the common good and with the "human element" in engineering always in view.

Underlying Pius's confident, productivist discourse was a theology of dominion: humanity rightly "subdues" creation for the sake of building a more just, fraternal society that anticipates the coming reign of God in its fullness. Indeed, the dominion mandate extended beyond the planet. Speaking to astronauts in 1956, the pope divined: "The Lord God who has placed in the heart of man an insatiable thirst for knowledge did not intend to limit his efforts at conquest when He said to him: 'subdue the earth' (Gen. 1:28) . . . He confided all creation to man."[10] On this view, history reaches its climax with humanity's redemption, while all other creatures here below and way beyond serve merely as backdrop.

In 1957, Pius delivered an address to machine-tool industry leaders in which he affirmed automation's relentless advance and liberating potential: "[T]oday machine tools have been perfected to a point where one can hope that an ever increasing number of workers will be freed who have until now been subjected to performance of purely material and monotonous tasks."[11] Pius also saw that cybernation eventually would replace many kinds of cognitive work. With automation yielding higher productivity and reducing the work week, both blue- and white-collar workers would face a new challenge—the right use of leisure—that required growth toward intellectual and spiritual maturity:

> Provided that man learns how to dominate his instincts and make good use of the ample means for developing his intellectual and spiritual life, there is nothing to fear from material progress. But should man yield to the temptation of a life of greater ease filled with more and more sensual pleasures, he will gain nothing from it except another kind of slavery and a certain moral decadence. We hope to the contrary that the

9. Pius XII, quoted in Haigerty, *Pope Pius XII and Technology*, viii.
10. Ibid., xvi.
11. Ibid., 21.

most profound needs of the soul will find their satisfaction in the greater amount of leisure time available because of modern machines.[12]

The pope's outlook on automation and the potential dangers and blessings of more leisure time take on new relevance in light of today's concerns over technological unemployment and the coming transition to post-jobs society (see chapter 8).

Church Teaching during the Aggiornamento

From 1961 to 1981, Catholic social teaching underwent significant development as the church sought to interpret and evaluate globalization processes and an emerging world risk society. With Vatican II the church renewed its own self-understanding and conception of its relationship to the world. While conservative elements clung to a fortress-church mentality, Catholic leaders and laity in many countries embraced Vatican II's call to openness, dialogue, and service. Catholic social teaching during these two decades was preoccupied with the superpower arms race and high levels of Cold-War military spending, the growing North-South gap and mass poverty in the Third World, and ongoing struggles for human rights and democracy in a post-Shoah and postcolonial context. The significance of the environmental crisis came into view in the early 1970s.

Methodologically, Catholic social teaching during the *aggiornamento* began to move away from its traditional reliance on natural law toward a biblical and theo-anthropological foundation for reflection on the church's place and role in modern, pluralistic societies. In *Gaudium et spes* and especially in John Paul II's social encyclicals, church teaching was grounded theologically in a christocentric anthropology.[13] Popes and bishops also drew upon both liberal and radical social theories in formulating their analyses of society, even as they retained the communitarian ethos and outlook associated with organic social theory. We can identify three tensions within the ecclesial documents, one created by the use of different social theories, another by an ambivalent response to techno-economic progress, and a third by a divided mind on methodology. As we'll see in chapter 4, these tensions reappear—albeit in somewhat different form—in more recent social teaching.

---

12. Ibid., 22.
13. Beigel, *Faith and Social Justice*, 82–85.

John XXIII's Vision of the Universal Common Good

Acknowledging a growing interdependence and "multiplication of social relationships" both within and between nations, John XXIII offered an optimistic appraisal of globalization in his social encyclical, *Mater et magistra* (1961), written at the dawn of the United Nation's first Development Decade. Although John "viewed the differentials in wealth among nations with great alarm as a serious obstacle endangering the possibilities for world peace," still he "assumed that the poor countries were only late-starters in the process of modernization and industrialization and would soon be developing their own economies."[14] Recognizing that the social question had become global, John urged leaders in the North to increase development assistance to the South while warning against neocolonial domination. The pope insisted that economic development must be balanced with social progress and conform to traditional norms of justice (MM #68–81), an "integral human development" position developed further by Paul VI in *Populorum progressio* (1968). Balanced, integral development also required attention to the rural crisis of underdevelopment, with modernizing state support for agriculture (e.g., price supports, access to credit, technical training) matched by community self-development; here the preferred model is family farms connected through cooperatives (MM #123–49).

In line with Pius XII, John exhibited a first-modern confidence in technological mastery of natural forces while also warning against misuses of technology. His dominion theology was wedded to a view of nature-as-inexhaustible-resource, an instrumental attitude typical of his time. Addressing the over-population issue, for example, the pope condemned "anti-life" technologies (i.e., birth control, sterilization) yet also endorsed the green revolution:

> The resources which God in His goodness and wisdom has implanted in Nature are well-nigh inexhaustible, and He has at the same time given man the intelligence to discover ways and means of exploiting these resources for his own advantage and his own livelihood. Hence, the real solution of the problem is not to be found in expedients which offend against the divinely established moral order and which attack human life at its very source, but in a renewed scientific and technical effort on man's part to deepen and extend his dominion over Nature. The progress of science and technology that has already been achieved opens up almost limitless horizons in this field. (MM #189)

---

14. Ellison, *Center Cannot Hold*, 79.

For John, amelioration of social problems associated with rapid population growth and expected resource scarcities could be achieved through economic growth, better technology, and distributive justice, not by population-control policies disrespectful of reproductive rights and human dignity. A decade later, the church withdrew its unqualified endorsement of techno-economic progress as the environmental movement reached critical mass and began to translate its demands into reform legislation in advanced industrial nations.

Moved by the Cuban missile crisis to address the dangers and costs of the Cold War, Pope John wrote a second major social encyclical, *Pacem in terris* (1963), in which he articulated a human rights ethic to specify the demands of justice on a global scale. Against one-sided readings, the pope maintained that civil/political and economic/social rights are of equal significance for the fulfillment of the human vocation within society; a lasting peace must be built on justice, which now must take the form of integral human development across the entire planet. Globalization required structures of international governance capable of protecting and promoting what John called the "universal common good," and thus he strongly endorsed the ongoing work of the UN and its agencies on behalf of world peace and development. In both social encyclicals, the pope adopted a gradualist view of progressive social change, with the principal social actors being responsible elites working cooperatively to establish a more just international order.

As Mary Hobgood points out, the social analyses of John XXIII contained elements compatible with both orthodox economic theory (i.e., neoclassical economics) and, to a lesser degree, radical social theory.[15] On the one hand, "John followed an orthodox analysis in his affirmation of private property and the wage system, and in his belief that genuine social betterment could occur through the application of western modes of technology to economic and political activity."[16] In keeping with liberal social theory, John held that nation-states should act as responsible agents of the universal common good by establishing structures of international governance

---

15. Hobgood, *Paradigms in Conflict*, 132–42. For Hobgood, liberal social theory includes a wide spectrum of views, from libertarian to progressive/social-democratic. Liberal social theorists typically "believe that the cultural and political spheres are relatively independent from the economic sphere" (22). Whatever the stripe, liberal social theory affirms a set of normative assumptions associated with neoclassical economics, including the merits of competition, pluralism, and functional hierarchies within economy and society. In essence, it explains and defends the structural features and core values of democratic capitalism. By contrast, radical social theory challenges the basic assumptions of liberal social theory and, for Hobgood at least, is informed by Marxist, anti-racist, and socialist feminist perspectives.

16. Ibid., 132.

to regulate global markets and capital flows while respecting the rights of private property and economic initiative. Within developed nations, John welcomed the construction of state-welfare systems as a proper instrument for increasing participation in social life and protecting human rights against market excesses. On the other hand, some aspects of John's social analyses resonated with the ideas and concerns of radical social theory, in particular his strong defense of economic/social rights and the just social use of property based on the universal destination of created goods and corresponding priority of meeting basic human needs. John also advanced Catholic social teaching on workplace democracy, arguing that "workers [ought to] become partners not only in the ownership of industrial, agricultural, and craft enterprises but in their control and management."[17] Finally, to check the power of transnational corporations, the pope reinterpreted the principle of subsidiarity so as to "emphasize the need for public authority to intervene when power was too great to be left in private hands without damage to the common good, especially the poor."[18]

Work and Leisure in a Technological Society

The Vatican II document *Gaudium et spes* (1965) expressed an open, positive view of technological society. *Gaudium* placed the human person—understood theologically as a person-in-community called to reflect the love and unity of a Trinitarian God (GS #24)—at the center of its social thought. The Vatican II council was eager to affirm all that is authentically human in the life of modern society, even as it called Christians to transform sinful, unjust conditions in a spirit of service and for the sake of contributing to God's coming reign. Here we consider the council's reflections on the meaning of work and leisure as well as the role of science and technology.

Whatever its character—physical or mental, menial or highly skilled—the bishops reaffirmed the received view that labor has a purpose beyond making a living and supporting a family. Work also provides opportunities for fellowship, service, and self-development. Through dignified labor a person "can exercise genuine charity and be a partner in the work of bringing divine creation to perfection" (GS #67). According to the council, certain norms follow from work's social meaning and redemptive significance: workers have both the duty and right to work, and society through its institutions has a general obligation to help them secure employment and pay them a living wage, i.e., one that enables the worker to "cultivate

17. Ibid., 136–37.
18. Ibid., 138.

worthily his own material, social, cultural, and spiritual life and that of his dependents" (GS #67). *Gaudium* also reflects an integral-humanist philosophy of the person as subject of work and as directed to other pursuits in life rather than "living to work":

> The entire process of productive work . . . must be adapted to the needs of the person and to his way of life, above all to his domestic life, especially in respect to mothers of families, always with due regard for sex and age. The opportunity . . . should be granted to workers to unfold their own abilities and personality through the performance of their work. Applying their time and strength to their employment with a due sense of responsibility, they should also all enjoy sufficient rest and leisure to cultivate their familial, cultural, social and religious life. They should also have the opportunity freely to develop the energies and potentialities which perhaps they cannot bring to much fruition in their professional work. (GS #67)

Well-designed workplaces prioritize the development of human personality and capabilities over tools of production and profits; they also abide by norms that keep work in its proper place, as it were. Long before work-life balance and meaningful work became major issues, *Gaudium* set forth clear guidelines based on Christian-humanist principles. In this way, conciliar teaching on work prefigures contemporary calls for a shift away from technology-centered automation, which tends to de-skill or eliminate workers while concentrating wealth, to the development and use of human-centered technologies at the service of more engaging, meaningful work.

Praise for modern science and technology was warranted—indeed its "wonders" amazed the bishops—provided such work contributed to a nobler humanity—a humanity attuned to and guided by moral and spiritual values—and did not lead moderns to deny or ignore a Christian revelation that transcends all scientific knowledge and technological prowess, however far-reaching and powerful it may become. Hence in *Gaudium* the council restated, albeit in less strident terms, the church's longstanding polemic against secular philosophies, including the materialist worldview often associated with modern scientific inquiry and technological development.

Gaudium reflected a neo-Thomist view of how technology and culture are related to salvific truth. The bishops reaffirmed Vatican I teaching on two orders of knowledge—faith and reason—where the light of reason finds its final meaning and fulfillment in the light revealed to faith. The work of natural reason, scientific and technological activity enjoys a real yet limited autonomy. Reason's limits are an ontological given: finitude and corruption from the Fall can be overcome only through the supernatural solution offered by God, a solution in which nature (reason) is completed and perfected by grace, as Aquinas puts it.

This premodern theological stance informed the council's critique of scientism and its position on the purpose of science and technology in modern society. Over and above the material benefits they confer and the broader sense of community they promote, scientific and technological pursuits "provide some preparation for the acceptance of the message of the Gospel, a preparation which can be animated by divine charity through Him Who has come to save the world" (GS #57). For Aquinas, classical philosophy serves as handmaiden to theology, and natural theology prepares the way for the full revelation of God in Christ. The bishops follow Thomas in holding that modern techno-scientific activity may open up vistas that lead the modern mind (by grace) to see and accept the saving truth of the gospel. By contrast, an anthropocentric and technocratic approach that seeks to grasp and control reality for the sake of self-transforming humanity according to some utopian ideal is fated to futility or at risk of becoming idolatrous and destructive. From Vatican I to Pius XII to *Gaudium* to *Laudato*, the church has based its critique of what Pope Francis calls the "dominant technocratic paradigm" (LS #101–2) on a theological anthropology that counter-poses an authentic religious humanism (or integral humanism) to all forms of modern anthropocentrism and secular humanism that deny or close themselves off to a transcendent horizon.

Paul VI and Integral Human Development

Addressing global development issues in *Populorum progressio* (1967), Paul VI evinced a more critical attitude toward modernization theory and practice. The pope recognized not only the failures of development policy to reverse mounting inequalities but also the legitimacy of growing demands for radical reform arising from the impoverished many. Against economistic views, the pope called for new policy initiatives to promote far-reaching structural change at both national and international levels based on a holistic, theological-ethical vision of human flourishing: "Development cannot be limited to mere economic growth. In order to be authentic, it must be complete: integral, that is, it has to promote the good of every man and of the whole man" (PP #4). The structural reforms and coordinated planning advocated by Paul were reflected during these years in the Non-Aligned Movement's proposal for a "new international economic order"—an alternative perspective reflected in the work of the UN Conference on Trade and Development, formed in 1963 to advance the interests of G-77 nations.

The heuristic notion of integral (or authentic) human development articulated by Paul VI in *Populorum* was informed by personalist philosophy (Maritain) and structured by an integral scale of values: vital, social, cultural, personal, and religious. In a good society, all of these values are protected, promoted, and passed on from one generation to the next; the integral scale ideally spirals "upward" from the satisfaction of each individual's *vital* (physical) needs to the (re)construction of a just *social* order, the flowering of *cultural* life, the expansion of opportunities for *personal* growth and character development, and ultimately to the free acceptance of *religious* truth and salvation in Christ (PP #21). This heuristic concept provides a broad ethical framework for evaluating whether ideologies, societal trends, government policies, business practices, and civil-society initiatives tend to support or undermine inclusive human flourishing. Where the trajectory of development is "from less human conditions to those which are more human" (PP #20), society is moving along a path of genuine social progress. The idea and ideal of integral human development anticipates current efforts in Bhutan, France, and elsewhere that attempt to gauge "gross national happiness" by using a broader set of indicators beyond GDP and the Physical Quality of Life Index.

Paul emphasized that all efforts directed toward authentic human development must be animated by a spirit of global solidarity and universal charity. Adopting a global perspective, the pope hoped that "peoples and nations would recognize their mutual interests in promoting development for all and, thereby, create a world community through dialogue and cooperation."[19] Both personal growth and community development contribute to the creation of what Paul referred to as a "civilization of love."

As with John XXIII, Paul's preferred consensus model of social change banks on the efficacy of appeals to reason and conscience. The principal change agents are responsible elites who come to see the wisdom of cooperating in the establishment of major structural reforms from the top down, if only on the basis of an enlightened self-interest concerned with maintaining systemic stability and avoiding the wrath of impoverished masses (PP #49). While not ruled out, strategies for social change by means of nonviolent confrontation or armed struggle initiated from below by the poor and oppressed are not encouraged. Violent revolution is at best a "last resort" under extreme conditions, an option fraught with its own perils (PP #33).[20] For both popes, a preference for change strategies based on consensus-building, cooperation, and the over-arching demands of the common good

19. Ellison, *Center Cannot Hold*, 84.
20. Dorr, *Option for the Poor*, 170–74.

can be traced to the residual influence of organic social theory as well as a core theological conviction: God and humanity are being reconciled in Christ, and hence integral to the church's mission is a fostering of peace and fraternity among all peoples. As a result, the transformative potential of Gandhian nonviolence as a method of social change does not gain a hearing.

Systemic Injustice and Calls for Liberation

Paul VI's apostolic letter *Octogesima adveniens* (1971) may be read as a response to calls for liberation coming from progressive forces within the Latin American church. The radical social analyses developed by liberation theologians assumed a revolutionary situation characterized by extreme social inequality, oppression, and massive human suffering—a systematic denial of human rights and values. A search for alternatives to modernization theory (Rostow) was prompted by the failure of various large-scale, top-down development schemes and land-reform programs as well as the rise of repressive national-security states in the region. Rejecting the historicist assumptions and ethnocentric biases of the modernization model, liberationists turned to dependency theory (a precursor of world systems theory) for a historical analysis of the colonial legacy and a structural analysis of the inequality and "institutionalized violence" generated by the global capitalist system.

In briefest sum, dependency theory views the social facts of pronounced class inequality, political repression, and uneven development as a function of the international division of labor. Through a variety of mechanisms (e.g., trade rules, loan policies, suppression of dissent), powerful countries at the "center" of the international economic system come to dominate countries on the "periphery" by exploiting the raw materials and cheap labor of the latter. This "development of underdevelopment" occurs not only in structured relations between rich and poor nations, but also within peripheral countries as a small, rentier elite benefits from foreign aid and investment in the process of exploiting rural areas and the urban working class. Lacking educational and work opportunities, and in the absence of state-sponsored social programs, the urban poor eke out a marginal existence within a precarious informal economy, and their numbers grow each year as displaced rural workers migrate to the slums ringing the cities. Dependency theory thus links the enrichment of powerful nations and local elites to the impoverishment of weaker countries and the working classes.[21]

---

21. See, e.g., Frank, *Capitalism and Underdevelopment*.

Political and pastoral responses to structural injustice in Latin America varied. For liberationists who rejected revolutionary struggle through guerrilla warfare, the most fitting response was a politics of radical reform aimed at promoting autonomous development within impoverished nations and constructing a new international economic order. Informed by liberation theology, many priests, nuns, and lay leaders set up base communities dedicated to Bible study and Freire's literacy method. Others became active in building and running schools and health clinics in poor communities, while still others focused on organizing students, workers, and dispossessed peasants. At Medellín in 1968 the Latin American bishops abandoned the New Christendom model of church-society relations, made extensive use of dependency theory in their social analyses, and adopted a pastoral strategy of actively defending human rights and supporting grassroots organizing among the poor for the sake of "authentic liberation."[22] Ten years later at Puebla, the bishops reaffirmed the church's "preferential option for the poor."

Although Paul VI echoed liberationist themes in *Octogesima* and recognized the "priority of the political," he remained critical of the use of Marxist theory and warned against its destructive implications. As in *Populorum*, Paul spoke forcefully of the need for a new international economic order. However, as Marvin Ellison observes, Paul "continued to address his appeals primarily to the rich nations to engage them in solving the problem of underdevelopment rather than calling upon the poor and oppressed to organize and free themselves from political and economic dependency and to become the artisans of their own development."[23] The consensus model for social change adopted in *Populorum* remained in place.

At the same time, Paul VI acknowledged the "aspiration to equality and the aspiration to participation" (OA #22) voiced by ordinary people demanding more say in decisions affecting their lives, including decision-making within "the social and political sphere" as well as the economic arena (OA #47). The pope also recognized the power of a *prophetic-utopian imagination* capable of critiquing injustice and envisioning a new society. Such an imagination engages not in ideal flights from historical reality, but rather "perceives in the present the disregarded possibility hidden within it" and seeks to direct it "towards a fresh future" (OA #37). Under oppressive conditions the risky practice of envisioning "concrete utopias" (Bloch) opens up avenues for change typically considered unthinkable.

According to Hobgood, Paul's social teaching mixed a radical critique of unjust power relations with analyses and policy prescriptions associated

---

22. Ellison, *Center Cannot Hold*, 31–43, 87–90.
23. Ibid., 85–86.

with the progressive/social-democratic strand of liberal social theory. At times the pope spoke of "abuses" in the global capitalist order that public authorities can and should remedy with bold reforms, e.g., establishment of a world fund devoted to development assistance for the poorest nations and financed by reductions in military spending. In other moments, however, Paul argued that "technocratic capitalism" was an inherently flawed, exploitative system incapable of meeting human needs. Drawing on dependency theory, Paul suggested that severe imbalances in international trade were not simply a function of differences in productivity levels, but stem rather from the exercise of market power by dominant countries. Paul also recognized that, given their immense power, transnational corporations were capable of "a new and abusive form of economic domination," since they were not accountable to individual nation-states "and therefore not subject to control from the point of view of the common good" (OA #44).[24]

Hobgood, Donal Dorr, and other commentators have noted the radical implications of a methodological shift in *Octogesima*. Paul opened the door for the use of a historically contextualized methodology among local/regional churches and acknowledged that, given the diversity of concrete situations across the world, the church "does not intervene to authenticate a given structure or to propose a ready-made model" (OA #42). By adopting an inductive approach, Paul sanctioned the possibility of plural approaches to reading the signs of the times and formulating action strategies, including a potentially legitimate use of radical social analysis as well as active Christian participation in liberation struggles (OA #4). Equally significant, in my view, were Paul's reflections on the challenges of urbanization and industrialization in which he introduced a concern for what John Paul II would later refer to as "human ecology" (*Centesimus annus*, #38) and articulated the church's initial response to the environmental crisis and an emerging world risk society.

Urbanization, Ecology, and the Rise of World Risk Society

Paul's discussion of urbanization in *Octogesima* highlighted the new social problems attending the inordinate growth of mega-cities across the globe and noted that, as a preferable alternative, "medium-sized towns" provided a more appropriate scale for meeting the challenges posed by rapid techno-economic development, population growth, and social dislocation. Within mega-cities, the restless energies of industrial capitalism generated new forms of alienation and a new underclass, while at the same time

24. Hobgood, *Catholic Social Teaching*, 152–54.

encouraging overconsumption and wastefulness among the privileged as well as indifference toward and discrimination against marginalized groups. "Within industrial society," a rapid and unplanned "urbanization upsets both the ways of life and the habitual structures of existence: the family, the neighborhood, and the very framework of the Christian community" (OA #11). Absent a healthy human ecology, integral human development turns into its opposite: social fragmentation, crime, low trust, and wasted potential. Hence the pope urged Christians to join with others "to remake . . . the social fabric whereby man may be able to develop the needs of his personality" (OA #11). Paul's concern for human ecology echoed several themes articulated by green-communitarian advocates of "livable cities" and "sustainable communities."

In *Octogesima*, Paul VI also recognized that industrial modernity was generating new hazards and threats that could not always be contained or controlled through technical expertise:

> [A]nother transformation is making itself felt, one which is the dramatic and unexpected consequence of human activity. Man is suddenly becoming aware that by an ill-considered exploitation of nature he risks destroying it and becoming in his turn the victim of this degradation. Not only is the material environment becoming a permanent menace—pollution and refuse, new illnesses and absolute destructive capacity—but the human framework is no longer under man's control, thus creating an environment for tomorrow which may well be intolerable. (OA #21)

The apocalyptic undertone of this passage signaled a growing awareness of the ongoing transition from industrial to world risk society. For Paul, the environmental crisis is not "out there" in nature, but rather is a "wide-ranging social problem which concerns the entire human family" (OA #21). The pope concluded that a new imperative of shared responsibility had arisen from such a situation: "The Christian must turn to these new perceptions in order to take on responsibility, together with the rest of men, for a destiny which from now on is shared by all" (OA #21). With these reflections, Paul introduced into the tradition not only a new issue—environmental degradation—but also a new horizon and context: world risk society. His concern for both natural and social ecologies has become a leitmotif of contemporary Catholic social teaching.

Paul's recognition of an emerging world risk society in *Octogesima* stands in stark contrast to earlier statements that virtually identify man's dominion over creation with the necessity and legitimacy of industrial progress. In *Populorum*, released only four years prior, Paul had stated that

> the Bible, from the first page on, teaches that the whole of creation is for man, that it is his responsibility to develop it by intelligent effort and by means of his labor to perfect it, so to speak, for his use . . . By persistent work and use of his intelligence man gradually wrests nature's secrets from her and finds a better application for her riches. As his self-mastery increases, he develops a taste for research and discovery, an ability to take a calculated risk, boldness in enterprises, generosity in what he does and a sense of responsibility. (PP #22, 25)

As with Pius XII and John XXIII, the confident, productivist discourse employed here by Paul reflected the assumptions of industrial modernity. The dominion theology found in *Populorum* was not yet tempered by ecological awareness and a dialectical view of techno-economic progress.

The ecological question raised by Paul in *Octogesima* was addressed by the world synod of bishops in "Justice in the World" (1971). A year before the Club of Rome released its influential report on "The Limits to Growth," the bishops noted the growing perception of a fragile, endangered earth and the sense of shared responsibility it evoked: "[P]eople are beginning to grasp a new and more radical dimension of unity; for they perceive that their resources, as well as the precious treasures of air and water—without which there cannot be life—and the small delicate biosphere of the whole complex of all life on earth, are not infinite, but on the contrary must be saved and preserved as a unique patrimony belonging to all human beings" (JW #8). Acknowledgment of the limits-to-growth thesis added a new twist to the bishops' analysis of neocolonial domination: overconsumption by the affluent, global North robbed present and future generations of their fundamental right to development by undermining the life-support systems upon which all depend. The imperative of shared responsibility was, at the same time, a requirement of global eco-justice, particularly for advanced-industrial countries:

> It is impossible to see what right the richer nations have to keep up their claim to increase their own material demands, if the consequence is either that others remain in misery or that the danger of destroying the very physical foundations of life on earth is precipitated. Those who are already rich are bound to accept a less material way of life, with less waste, in order to avoid the destruction of the heritage which they are obliged by

absolute justice to share with all other members of the human race. (JW #70)

This passage reflected a just-sustainability perspective advocated by green communitarians, and the document as a whole closely linked faith with liberating social action (a connection reaffirmed recently by Francis in *Evangelii Gaudium*). Nonetheless, on the question of social-change strategy the bishops continued to follow a mediatory, consensus-building approach focused on moral appeals and thus shied away from direct calls for the poor and oppressed to confront the powers that be with demands for systemic change.

John Paul II's Early Social Teaching

The ongoing transition from industrial to world risk society figures prominently in John Paul II's early social teaching. In *Redemptor hominis* (1979), the pope spoke eloquently and at length about humankind's growing awareness of living under the "power of threat" (Beck). Alienation occurs not only when workers no longer own the means of production and are forced to create commodities that bear no stamp of their own creativity, but also when advanced industrial production generates unintended and potentially uncontrollable side-effects that imperil civilization:

> The man of today seems ever to be under threat from what he produces, that is to say from the result of the work of his hands and, even more so, of the work of his intellect and the tendencies of his will . . . Man therefore lives increasingly in fear. He is afraid that what he produces—not all of it, of course, or even most of it, but part of it and precisely that part that contains a special share of his genus and initiative—can radically turn against himself for an unimaginable self-destruction, compared with which all the cataclysms and catastrophes of history known to us seem to fade away. (RH #15)

In John Paul's judgment, the emergence of world risk society and its nuclear, genetic, chemical, and ecological mega-hazards was not simply one sign of the times among others. Rather, it constituted "the main chapter of the drama of present-day human existence in its broadest and most universal dimension" (RH #15).

For John Paul, runaway technology was not a structural feature of modernization. Rather, the dark side of modern progress was symptomatic of an ethical failing and revealed a situation of social sin: industrial moderns have distorted the dominion mandate (Gen 1:27–28), which calls for the exercise of responsible stewardship, by creating a high-throughput economy and consumer/throwaway culture that systematically over-exploits the earth in the relentless search for greater power, profit, and affluence. Left uncontrolled, modern techno-economic progress degrades the natural environment and alienates humanity from nature. The pope observed that industrialized humanity "often seems to see no other meaning in his natural environment than what serves for immediate use and consumption. Yet it was the creator's will that man should communicate with nature as an intelligent and noble 'master' and 'guardian,' and not as a heedless 'exploiter' and 'destroyer'" (RH #15). While he echoed the familiar Romantic complaint that incessant getting and spending dulls our capacity for an aesthetic appreciation of nature, John Paul did not advocate a return to a simple, human-scale society closely attuned to nature's rhythms. Instead, a proper use of resources in a technologically advanced world "demands rational and honest planning" respectful of planetary boundaries. Yet absent a moral-spiritual commitment to responsible stewardship, the value of rapid technological development would prove ambiguous at best.

John Paul II's *Laborem exercens* (1981) is best known for its championing of the human person as the subject of work and not merely a human resource at the disposal of capital. According to Patricia Lamoureux, "the encyclical proffers a vision of work as 'transformative vocation.' That is, work is a way men and women collaborate in the ongoing creation and re-creation of self, the workplace, and the world."[25] As noted above, the bishops in *Gaudium* held that work was integral to the human vocation; persons are called "to be a partner in the work of bringing God's creation to perfection" (GS #67). With *Laborem*, John Paul extended and deepened conciliar teaching on work.

From the perspective of world risk society theory, though, *Laborem* was a document written from within the horizon of early modernity, since it presumed that humans remain the masters of technology and never its victims, provided that relations between labor and capital are ordered rightly. Technology is an ally, an instrument at the service of workers, who reach their full potential through meaningful, productive labor as they cooperate to "subdue the earth" for the sake of authentic human development. However, where social injustice distorts economic and political systems, technology may become an "enemy" of labor, "as when the mechanization of work 'supplants' him, taking away all personal satisfaction and the incentive to creativity and responsibility, when it deprives many workers of their

25. Lamoureux, "Commentary," 408.

previous employment, or when, through exalting the machine, it reduces man to the status of its slave" (LE #5). On this reading, modern technologies become alienating and destructive of natural and social ecologies only when capital wrongly takes priority over labor. As with the environmental crisis, technology's dehumanizing impacts stem from disordered social relations and are not a systemic feature of industrial modernity per se.

In *Laborem*, John Paul argued that a condition of social sin arose from the ethical failings of early capitalists, who overturned the moral order by organizing the capital they controlled in ways antithetical to dignified human labor. "This disruption of the moral order . . . led to a false economic theory, which in turn made the opposition between capital and labor universal in modern society."[26] The roots of social injustice are found not in the contradictions of industrial capitalism as an inherently exploitative system, as Paul VI had suggested, but rather in the ideological error of "economism," which led capitalists to continue recreating economic arrangements that overturned the principle of the priority of labor over capital. Now globalized, the social question is *the* moral scandal of the modern age and must be overcome through workers' ethical (not class) struggle for social justice, pursued in solidarity with the poor for the sake of liberating every marginalized group in society from dehumanizing conditions.

According to John Paul, establishing the priority of labor over capital as the operative principle within a market economy required new institutional forms such as co-ownership and co-management of enterprises as well as responsible, democratically controlled economic planning. Once the principle of the priority of labor over capital was institutionalized, the pope hoped that cooperating managers and workers would act as responsible stewards, using appropriate technologies and natural resources for human development while also respecting the "integrity of nature."

In John Paul II's early social teaching, then, the early-modern confidence of *Laborem*—where technology-as-instrument serves the cause of human mastery over nature—appears to be in tension with the dialectical, late-modern account of techno-economic progress found in *Redemptor*. On the one hand, the pope recognized that modern industrial processes systematically yield mega-hazards that pose existential risks for humanity. On the other hand, any form of economic or technological determinism is rejected. Ecological degradation and oppressive technologies are manifestations of social sin, which always stems from personal sin. Hence John Paul's theo-anthropological response to systemic injustice was to call for the conversion of persons away from selfish individualism and consumerism

---

26. Baum, *Priority of Labor*, 57.

toward a universal solidarity grounded in love of God and neighbor. Such a conversion would bear fruit in a renewed commitment to social justice and responsible stewardship.

Methodologically, John Paul II's approach to what had now become the eco-social question may be criticized on two counts.[27] First, it suffered from the lack of a contemporary social theory. As Robert Simons notes, in John Paul's social encyclicals the task of analyzing the current state of affairs with the aid of modern social theory is cut short by a premature and "insufficiently critical recourse to religious language, theological concepts and Scripture."[28] In *Laborem*, for example, the pope problematically linked the human phenomenon of "toil" (Gen 3: 17) to the sufferings of Christ on the Cross. As Lamoureux observes, "to equate the 'toil' of work with the sufferings of Christ without attending to its causes in particular cases can lead one to overlook social injustice; and . . . John Paul's 'co-redemptive' view of the toil of human labor seems to imply that the more unpleasant work is, the more fully it is a participation in Christ's sufferings, and, paradoxically, the more redemptive it is."[29] In this case, John Paul's scriptural theology undermines his tireless defense of exploited, marginalized, and enslaved workers across the world.

Lamoureux's criticism points to a second, broader critique of John Paul's methodology. In his social teachings, the pope assumed a unitary body of magisterial social doctrine—a received set of universal, absolute norms and timeless social principles that transcend the contingencies of culture and history—which the faithful can then "apply" to specific circumstances. As the just-noted example from *Laborem* illustrates, this deductive and ahistorical methodology runs the risk of obscuring or even distorting the specific historical context within which Christian communities seek to discern God's will through reflection on their lived experience. Such reflection involves a hermeneutic process in which life stories and community experiences are illuminated by social theory as well as by prayerful reflection on Scripture and church teaching. In this respect, the methodological stance of John Paul II represented a retreat from the inductive, historically conscious approach advanced by Paul VI in *Octogesima*.[30]

---

27. The methodological critique presented here applies to the entire corpus of John Paul II's social teaching.
28. Simons, *Competing Gospels*, 118.
29. Lamoureux, "Commentary," 406.
30. Elsbernd, "*Octogesima Adveniens*?"

Three Tensions and Two Imaginations

To sum up, from 1891 to 1981 the church's response to rapid advances in technology, globalization, and the rise of world risk society was informed by its own evolving vision of integral human development, a social-ethical vision based on a theological anthropology, structured according to an integral scale of values, and specified by a comprehensive view of universal human rights. Influenced by the renewal of scriptural theology, there was a growing conviction among church leaders that social sin exists in multiple forms and at many levels, that working for social justice is a constitutive dimension of Christian faith, and that the church must make a gospel-inspired preferential option for the poor. Recognizing that industrial modernity was generating existential risks to humanity, the church issued a call to shared responsibility and linked it to demands for global eco-justice.

My analysis has drawn attention to three tensions within Catholic social teaching during the postwar period. The first was created by a mixed use of organic, liberal, and radical social theories. Here a turn to dependency theory and increasingly sharp critique of structural injustice sat uneasily with a reticence to depart from a consensus-building approach to social change appealing primarily to elites. The closest the church came to supporting change from below was John Paul II's careful delineation of an ethical (not class) struggle undertaken by workers to reestablish in practice the principle of the priority of labor over capital in economic life.

A second tension resulted from ambivalence toward industrial modernity and its boomerang effects. On the one hand, Catholic social teaching exhibited an early-modern confidence in the material progress wrought by modern science and technology, a confidence validated theologically by a productivist reading of the dominion mandate. On the other hand, with the release of *Octogesima* and "Justice in the World" in 1971 the church began to address the environmental crisis and develop a more dialectical, rather than progressivist, reading of techno-economic development.

A third tension arose from different methodological approaches. John Paul II's deductive, ahistorical method begins with Scripture-based doctrine and anthropological truths derived from personalist philosophy, which are then applied as critical correctives to a social reality distorted by sin. By contrast, Paul VI's inductive, pluralist method begins with local Christian communities seeking to make sense of a complicated, conflict-ridden reality with the aid of social theory and to discern God's will in light of their lived experience and a contextualized reading of Scripture and church teaching. In the next chapter, I show how these tensions reappear under a different guise as Catholic social teaching engages more deeply with the sustainable-development debate and weighs in on the emerging debate between proactionaries and precautionaries over the promise and peril of new technologies.

Finally, I noted how two forms of imagination—communitarian and prophetic-utopian—inspired and informed modern Catholic social teaching, though in quite different ways. A Catholic communitarian ethos and imagination has underwritten principled opposition to the ideological excesses of liberalism and socialism since the nineteenth century, while more recently social-justice and liberationist elements within the church have exercised a prophetic-utopian imagination in their quest for systemic change. While these two imaginations often are viewed as existing in tension with one another, or even seen as diametrically opposed, I argue in Part II that Christian praxis and moral leadership going forward will depend crucially on finding ways to combine a communitarian ethos with the generative power of prophetic-utopian imagination.[31]

---

31. Cf. Kearney, *Wake of Imagination*, 366–71. Kearney argues that what he calls a "poetical imagination" committed "to the invention of an alternative *social* project" (370) complements an "ethical imagination" attentive in Levinasian fashion to the concrete, suffering other in our midst; hence the need to develop and exercise a "poetical-ethical imagination." On my reading, we can speak analogously of the need for socially-conscious Christians to combine a renewed (and chastened) communitarian ethos with a prophetic-utopian imagination, or what I call the practice of envisioning and pursuing the adjacent possible of sustainable abundance for all (see chaps. 6 and 7).

4

Church Teaching on Sustainable Development and Technology (1987–2004)

From the late 1980s into the early 2000s, the debate over sustainable development became more urgent and more contested. Despite growing concerns over climate change, loss of biodiversity, and other environmental problems, the limits-to-growth argument was sidelined as an aggressive neoliberal agenda bent on global economic expansion and integration gained ascendance, e.g., through the newly established World Trade Organization. At the same time, ecological modernizers made headway in the EU and at the 1992 Rio Summit on Environment and Development with legal establishment of the precautionary principle, while in some business circles a "corporate sustainability" movement took hold.[1] For their part, green communitarians grew increasingly alarmed by widening socioeconomic inequality and mounting environmental stress. The Great Acceleration had pushed the planet "beyond the limits" of its carrying capacity, Club of Rome researchers again argued, with global collapse to follow overshoot if calls for major changes in lifestyles, practices, and policies were ignored.[2] Meanwhile, as structural-adjustment policies turned the screws on highly-indebted poor countries, "IMF riots" broke out in one city after another. In response, greens took to the streets in Seattle, Quebec, and elsewhere to protest against neoliberal policies that, in their view, granted mobile capital an unfair advantage over workers, national governments, and local communities.

Attentive to realities on the ground through the work of Caritas International and other social ministries, the church deepened its engagement with

1. See Vogel, *Market for Virtue*.
2. Meadows et al., *Beyond the Limits*.

the sustainable-development debate as the new millennium approached. Catholic social teaching reaffirmed its sharp critique of neoliberalism, made overtures toward both ecological modernizers and green communitarians, and continued to emphasize a global eco-justice perspective sensitive to the impact of environmental degradation on the poor.

Not surprisingly, the three tensions traced out in chapter 3 resurfaced in ecclesial documents, albeit in somewhat different form. First, just as postwar Catholic social teaching drew upon different social theories, papal and episcopal statements during the post-Cold War period appropriated insights from two distinct sustainable-development policy narratives: the pollution-prevention-pays storyline of eco-modernizers and green just-sustainability vision. Second, the church remained ambivalent toward industrial progress, oscillating between early-modern and late-modern perspectives. And third, different methodologies were employed; while papal teaching exhibited a deductive approach, episcopal statements were more contextual and inductive in method.

Church teaching in these years continued to reflect a communitarian ethos and occasionally exercise a prophetic-utopian imagination (e.g., the Jubilee 2000 campaign, which called for developing country debt cancellation). Drawing on premodern traditions of natural theology and Thomist ontology, John Paul II's statements on ecology emphasized the beauty, harmony, and order of all things brought into being and sustained by a loving Creator. Increasingly, John Paul's writings and sermons also included language evoking an *arcadian imagination*, i.e., language reminiscent of a Romantic view of nature-as-source as well as early-modern ecological ideas regarding the "balance of nature." For the pope and other modern Europeans, the arcadian imagination hearkens back to a "timeless" life-world—the idyllic landscapes and rural cultures of preindustrial England and the Continent—and expresses a certain moral sensibility associated with them, one marked by a sense of proportion, serenity, and fecundity. In this imaginary, village communities coexist peacefully with a bountiful nature, a local agricultural economy centered on the household operates within the much larger natural economy according to time-honored traditions, and God's handiwork is everywhere to be seen.[3]

Just as a Romantic conservatism rooted in medieval nostalgia served as one source among several for a still-resonant Catholic communitarian ethos and imagination, so too a Romantic vision of nature-as-source grounded in rural nostalgia generated an arcadian imagination that continues to inspire contemporary conservation efforts and a renewal of the stewardship ethic.

3. Worster, *Nature's Economy*, chap. 1.

The arcadian vision of a fecund, pacific nature full of beauty and in balance finds expression in John Paul II's references to the aesthetic value of creation and calls to respect a created order that is endowed with its own integrity.

While the church during this time maintained a traditional theology of dominion and continued to link it with modern techno-economic progress, a shift in emphasis occurred in its interpretation of the dominion mandate. The language of mastery—e.g., "subduing the earth" (Gen 1:28)—became less prominent by comparison with that of responsible stewardship—e.g., "tilling and keeping the garden of God" (Gen 2:15). In the *Compendium of the Social Doctrine of the Church* (2004) we read: "At the summit of this creation, which 'was very good' (Gen. 1:31), God placed man. Only man and woman, among all creatures, were made by God 'in his own image' (Gen 1:27). The Lord entrusted all of creation to their responsibility, charging them to care for its harmony and development (cf. Gen 1:26–30)" (#451). Here and elsewhere, a renewed theology of creation served to ground a stewardship ethic focused on caring for both natural and social ecologies. Nonetheless, the church continued to express an early-modern confidence in techno-scientific progress even as it voiced late-modern concerns about its dark side. Ambivalence toward industrial modernity led the magisterium in these years to adopt a weak precautionary stance on issues of technological and environmental risk.

After critically reviewing John Paul's later social teaching, I consider a remarkable pastoral message issued by the bishops of Appalachia in 1995. The Appalachian document exemplifies how a good deal of national/regional episcopal teaching shifted (in this case, significantly) during this period toward adoption of a just-sustainability perspective. A later section analyzes and critiques the *Compendium*'s chapter on environment and technology, which represents the church's initial response to an emerging proactionary-precautionary debate.

Deeper Engagement with the Sustainable-Development Debate

As noted last chapter, the church initially engaged issues of sustainability amidst growing ecological concern in the 1960s. The environmental crisis prompted Paul VI and the world's bishops to address the issue in 1971, a year after the first Earth Day and a year before the first major international conference on environment and development held in Stockholm. In "Justice in the World," the bishops were persuaded by the limits-to-growth thesis and linked it to their structural analysis of global inequality. Mounting environmental problems aggravated an already intolerable situation of systemic injustice, where the affluence of advanced-industrial nations came at the expense of impoverished (not merely undeveloped) countries in the Third World. For the bishops, a dual-awareness of planetary boundaries and grossly uneven development gave rise to demands for global eco-justice, which was framed as an issue of redistribution requiring changes in individual lifestyle and major policy reforms at national and international levels.

During the late 1980s—a time in which acid rain and ozone depletion dominated the environmental agenda while awareness of global warming began to go mainstream—the church deepened its engagement with the debate over sustainable development. John Paul II's *Sollicitudo rei socialis* was released in 1987, the same year as the Brundtland Report, a charter document for international efforts to promote sustainable development. Written to commemorate Paul VI's *Populorum progressio*, John Paul's social encyclical focused on how continuing Cold War tensions contributed to growing inequality and poverty across the world. The moral scandal of underdevelopment in the global South, the pope lamented, was matched by unacceptable forms of "superdevelopment"—i.e., patterns of overconsumption and waste—in the global North.

John Paul's outlook on globalization was not entirely negative, however. Among the positive signs of the times was rising ecological concern, "a greater realization of the limits of available resources, and of the need to respect the integrity and cycles of nature and to take them into account when planning for development" (SRS #26). The pope emphasized that authentic human development must be guided by a stewardship ethic, which enjoins (1) moral respect for both other living beings and the ecosystems that sustain them, (2) conservation of nonrenewable resources, and (3) concern for environmental quality and health in urban areas. States had a responsibility to establish strict environmental standards and comprehensive land-use planning standards (SRS #34).

More so than Paul VI, John Paul incorporated ethical concern for the natural environment into the church's vision of integral human development. Some environmentalist critics misread Catholic social teaching (and Christian theology generally) as irredeemably anthropocentric,[4] but the classic Christian view of creation and the human place/role in it is better characterized as theocentric and "anthropo-apical" (Rolston). The dominion mandate cannot be separated from the duty of responsible stewardship: all God-given life deserves respect and care, and humans are called to be

---

4. White, "Historical Roots"; see also Noble, *Religion of Technology*.

guardians and moral overseers of God's good creation, not its ruthless exploiters. This classic position was re-emphasized by John Paul, who spoke often of the need to respect the integrity of the created order, or what his successor Benedict XVI referred to as the "grammar of creation."

Consonant with *Laborem exercens*, which viewed social injustice as rooted ultimately in personal sin, *Sollicitudo* traced the root cause of poverty and environmental degradation back to individual moral failings that distort the integral scale of values. Wealth and power become ends in themselves rather than limited goods meant to serve the higher ends of morality, culture, and religion. Disordered souls contribute to a building up of sinful, dysfunctional structures that undermine natural and social ecologies alike. The tragic consequences of social sin reflect a distortion of the dominion mandate and lack of a stewardship ethic (SRS #29). For John Paul, such a situation called for moral-religious conversion, made good through concrete acts of solidarity. Accordingly, John Paul echoed Paul VI's plea for redistributive measures at national and international levels in a spirit of universal solidarity (SRS #39).

While John Paul steered away from any sustained discussion of specific sustainable-development pathways in *Sollicitudo*, the neoliberal sustainable-growth agenda was rejected on ethical grounds as too economistic and individualistic. Both here and in *Laborem*, the pope adhered to the church's longstanding critique of liberal capitalism as an ideological error that has contributed to overdeveloped consumer societies in the global North as well as mass poverty and underdevelopment in the global South. Wherever neoliberal ideology distorts economic and political systems, both personal and social reforms are necessary in order to protect human rights and promote the common good. The new emphasis in *Sollicitudo* was that reforms ought to incorporate a moral, and not simply prudential, respect for animals, species, and ecosystems, and that development projects and policies should recognize and safeguard the human right to a safe, healthy environment. While global justice demanded an equitable sharing of the earth's bounty within and between nations based on the universal destination of created goods, responsible stewardship required careful management of natural resources through regulatory regimes, land-use planning, and the like. In practice, this would issue in state policies at multiple levels more compatible with ecological modernization than with neoliberalism.

On the question of social-change strategy, John Paul II leaned toward a reformist, top-down approach in *Sollicitudo*, though bottom-up approaches were not ignored. For example, the pope's moral appeal to elites for reforms of international systems of trade and finance (SRS #43) was matched by a call for poor nations to initiate their own plans for regional economic cooperation and development (SRS # 44–45). Overall, the emphasis fell on elite responsibilities, though the pope did voice support for impoverished communities demanding human rights through acts of nonviolent protest (SRS #39). When compared with *Laborem*, however, ethical norms typically associated with radical social theory such as equality and participation were not stressed as much, and both solidarity and the preferential option for the poor were associated more with individual expressions of Christian charity than with worker struggles for social justice.[5] Despite the trenchant critique of superdevelopment and structures of sin, *Sollicitudo* offered relatively little support for the kind of change from below advocated by green communitarians. That said, the document ends with a clear affirmation of and call to social action grounded in Christian hope for a societal transformation that prefigures the fullness of salvation in the future. The sorry state of international development notwithstanding, "the Church must strongly affirm the possibility of overcoming the obstacles which, by excess or by defect, stand in the way of development. And she must affirm her confidence in a true liberation. Ultimately, this confidence and this possibility are based on the Church's awareness of the divine promise guaranteeing that our present history does not remain closed in upon itself but is open to the Kingdom of God" (SRS #47).

A Lean toward Eco-Modernizers, a Nod to Greens

Read in light of the sustainable-development debate, John Paul's 1990 World Day of Peace message, "The Ecological Crisis: A Common Responsibility," was notable in three respects. First, the papal message provided more explicit, detailed attention to and support for an eco-modernizing agenda than is found in *Sollicitudo* or the pope's 1991 social encyclical *Centesimus annus*, while it also included green perspectives. Second, in line with *Redemptor hominis*, the pope's message expressed a dialectical view of techno-economic progress and addressed environmental issues from within the horizon of world risk society. Third, the ethical norm of respect for life was extended to include future generations. Noting the seriousness of environmental injustice, John Paul also acknowledged calls for an updated Charter of Human Rights to ensure the "right to a safe environment."

As in *Sollicitudo*, "The Ecological Crisis" viewed environmental problems as symptomatic of a deeper moral-religious crisis requiring conversion. John Paul emphasized that persons living in overdeveloped consumer

---

5. Hobgood, *Catholic Social Teaching*, 193–94.

societies must accept responsibility for making changes in lifestyle necessary for a just, sustainable future. The exercise of personal virtues such as frugality complements social action embodying the virtue of solidarity. In its institutional dimension, the social virtue of solidarity required new forms of cooperation among decision-makers at national and international levels.

For John Paul, a religiously based stewardship ethic provides the necessary moral foundation for sustainable-development policies along the lines of progressive ecological modernization. For instance, the pope advocated for a *"more internationally coordinated approach to the management of the earth's goods"* (#9) as well as strong national policies aimed at protecting environmental and human health. Advanced-industrial nations must take the lead in applying strict environmental standards within their own borders before expecting countries in the global South to follow suit. At the same time, John Paul stated that industrializing nations "are not morally free to repeat the errors made in the past by others, and recklessly continue to damage the environment through industrial pollutants, radical deforestation, or unlimited exploitation of non-renewable resources" (#10).

"The Ecological Crisis" also contained passages resonant with the just-sustainability vision of green communitarians. Befitting the occasion, John Paul made a clear connection between aspirations for peace and respectful treatment of both vulnerable human beings and nature: "*[N]o peaceful society can afford to neglect either respect for life or the fact that there is an integrity to creation*" (#7). As Vandana Shiva and other green-communitarian voices insist, the growth imperative of global capitalism amounts to a "war on nature and the poor" when, for example, massive extraction of natural resources (e.g., conflict minerals) and industrial-monoculture farming lead directly to slave labor, suicide, degraded soils, and the impoverishment of millions in sub-Saharan Africa, rural India, Amazonia, and elsewhere.

Echoing a just-sustainability perspective informed by structural analysis, John Paul noted as well how the search for sustainable development would be in vain unless "*the structural forms of poverty that exist throughout the world*" (#11) were addressed forthrightly by all responsible parties. Unjust patterns of land distribution in many countries contribute to a vicious cycle of environmental degradation and poverty for the rural poor, resulting in migration to already overburdened urban areas. Empowerment of the poor "will require a courageous reform of structures, as well as new ways of relating among peoples and States" (#11).

Heirs to the Romantic tradition and its arcadian imagination, many greens draw inspiration from wild nature and, despite the dawn of the Anthropocene, still find aesthetic and therapeutic value in direct, sensuous contact with a natural world that precedes and exceeds the human presence.

Accordingly, they seek to create built environments that mirror nature's beauty, respect its rhythms and contours, and remain at modest human scale. Leave grandeur to the Grand Canyon, they argue, and build accordingly. Similarly, John Paul called in "The Ecological Crisis" and elsewhere for a renewed appreciation of the aesthetic value of creation and practice of good urban planning as an antidote to modern forms of alienation:

> Our very contact with nature has a deep restorative power; contemplation of its magnificence imparts peace and serenity. The Bible speaks again and again of the goodness and beauty of creation, which is called to glorify God . . . More difficult perhaps, but no less profound, is the contemplation of the works of human ingenuity. Even cities can have a beauty all their own, one that ought to motivate people to care for their surroundings. Good urban planning is an important part of environmental protection, and respect for the natural contours of the land is an indispensable prerequisite for ecologically sound development. The relationship between a good aesthetic education and the maintenance of a healthy environment cannot be overlooked. (#14)

While the reference to Scripture makes clear that John Paul's nature aesthetic was grounded in ancient Christian traditions, the language of nature's "deep restorative power" and its capacity to impart "peace and serenity" echoed Wordsworth and other Romantics. The connections made between aesthetics and "designing with nature" stemmed as much from an arcadian imagination as from sacramental vision.

Continuing Ambivalence toward Progress

A late-modern awareness of techno-economic development's dark side permeated the pope's World Day of Peace message. John Paul identified "the *indiscriminate application* of advances in science and technology" as a major cause of the environmental crisis and stated that such problems as acid rain and climate change had "now reached crisis proportions as a consequence of industrial growth, massive urban concentrations and vastly increased energy needs" (#6). In other words, as Ulrich Beck also observed in these years,[6] it is the *success* of industrial modernization that systematically generates the mega-hazards now threatening human and environmental health on a global scale. Furthermore, the pope implied the need to invoke the

---

6. Beck's seminal work, *Risk Society: Towards a New Modernity*, was originally published in Germany in 1986, the same year as the Chernobyl nuclear disaster.

precautionary principle to safeguard natural and social ecologies from the potential dangers of bio-technological experimentation: "We are not yet in a position to assess the biological disturbance that could result from indiscriminate genetic manipulation and from the unscrupulous development of new forms of plant and animal life, to say nothing of unacceptable experimentation regarding the origins of life itself" (#7). This cautious approach to genetic engineering indicated an ongoing shift within Catholic social teaching from an early-modern enthusiasm for scientific progress to the recognition of living with a welter of "manufactured uncertainties" (Giddens) in world risk society.

However, Catholic social teaching during the 1990s remained ambivalent toward industrial modernity, at times expressing an early-modern confidence in technocratic capabilities while in other moments voicing late-modern concerns about high-risk technologies. Consider, for example, a Vatican statement from 1998 on the prospects for peaceful and safe uses of nuclear energy. In his address to the International Atomic Energy Agency, Monsignor Diarmund Martin spoke of the need to develop a "global nuclear safety culture."[7] Mons. Martin appeared confident that nuclear energy could be deployed and managed safely worldwide and suggested that, as a matter of global sustainable-development strategy, nuclear energy might play a role in mitigating climate change: "In the case of commercial energy services," the monsignor stated, "an impartial and objective judgment is still needed in order to achieve a fair balance between the long-term risks and the potential contribution of nuclear energy to a rapid and consistent diminution of carbon dioxide and other greenhouse gases."[8] From the perspective of world risk society theory, what is interesting about Mons. Martin's comment is not the issue, i.e., the difficult tradeoffs associated with energy options. Rather, it is the early-modern assumption that it is possible to render "*an impartial and objective judgment*" (my emphasis) regarding the "long-term risks" of nuclear energy relative to other options and scenarios, e.g., the risk of runaway climate change if greenhouse-gas emissions go unchecked. However, as Beck, Nassim Taleb, and others argue, reliance upon current risk-assessment methods is problematic, given the predilection of risk analysts to make rosy assumptions about the calculability of risks even when dealing with large, complex systems. I return to these issues below as well as in chapter 9 with a case study on energy options in south Florida.

In *Centesimus annus* (1991), John Paul II reflected on the downfall of communism and certain aspects of an accelerating globalization process.

---

7. Martin, "Nuclear Energy."
8. Ibid.

While the pope's positive evaluation of the modern business economy and critique of bureaucratic welfare systems was congenial to neoliberalism, he warned that the demise of state socialism must not be interpreted as a vindication of free-market ideology. The encyclical specifically rejected structural-adjustment policies favored by neoliberals and called for least-developed-country debt relief and even debt cancellation.

John Paul sought in *Centesimus* to correct the errors of neoliberalism by setting out broad ethical guidelines for the development of an alternative "society of free work, of enterprise and of participation" in which meeting basic needs takes priority over maximizing profits (#35). These guidelines included: (1) the right to economic initiative and duty of self-development, (2) the utility of market mechanisms when kept within moral-cultural limits and guided by state regulations that safeguard the common good, (3) the social obligations attached to private property, (4) the principle of subsidiarity, (5) the principled rejection of consumerism and commitment to "life-styles in which the quest for truth, beauty, goodness and communion with others for the sake of common growth are the factors which determine consumer choices, savings and investments" (CA #37), and (6) the renewal of a responsible stewardship directed toward care of both natural and social ecologies. Although the pope accorded comparatively little attention to a structural analysis of global capitalism and seemed to distance official church teaching from a liberationist perspective, he did speak approvingly of postwar social democracies that have balanced economic growth with social inclusion.

For John Paul II, an adequate response to the eco-social question demands both a change of hearts and a restructuring of institutions; conversion and spiritual growth is authenticated in concrete forms of solidarity, which include both selfless service on behalf of the common good and structural reforms that address systemic injustice. While few would argue with the proposition that both personal and social change is necessary to address the eco-social question, certain tensions remained within John Paul's social teaching. For example, his preferred consensus model for social change—a model heavily dependent on the efficacy of calls to conversion and moral appeals to elites—sat uneasily with his analysis of social sin and its devastating effects on marginalized, exploited populations. Granted, the pope spoke in *Laborem* and elsewhere of the need for and legitimacy of a workers' ethical struggle for justice. Nonetheless, his social teaching lacked a well-developed social analysis and theory of social change capable of funding a long-term, strategic view of systemic change. On many public occasions the pope boldly spoke out against injustice, yet his underlying commitment to a communitarian ethos and ecclesial unity as well as his Augustinian sense of pervasive human sinfulness and deep appreciation for the ironies of history rendered him wary of liberationist models of radical social change. Unlike Paul VI, John Paul did not speak of a utopian imagination that discerns hidden possibilities for social transformation within the present.

In relation to the sustainable-development debate, John Paul's social teaching reflected a tension between support for an eco-modernizing approach, on the one hand, and sympathy for green-communitarian ideas, on the other. With the ecological modernizers, the pope envisioned an expanding global economy in which G-77 nations would become integrated into world markets on fair terms through the cooperative efforts of G-7 elites; both moral duty and enlightened self-interest would lead the global North to help the global South undergo ecological modernization (e.g., through technology transfers) as quickly as prevailing conditions allowed. The hope was that a political consensus could be forged on the need for major policy reforms that would establish governance regimes at national and international levels capable of protecting human rights, including the right to a safe environment.

At the same time, John Paul championed eco-justice concerns dear to greens, e.g., in his repeated calls for defense of indigenous peoples' rights and agrarian reform in countries plagued by corrupt latifundia systems and neocolonial exploitation of the Amazon and other rural-wild regions across the planet.[9] In important respects the pope's outlook—and Catholic social teaching as a whole—shares an affinity with green communitarian thought. Both place a high value on human-scale communities within which natural and social ecologies may flourish, and both advocate the democratization of wealth and economic opportunity. Several core principles of Catholic social teaching—including solidarity, subsidiarity, pluralism, equality, and participation—are central to a green-communitarian vision of and program for just sustainability.[10] The section to follow provides a notable example of how this affinity shaped episcopal social teaching on the challenges of social justice and sustainability in Appalachia.

A Shift toward Just Sustainability in Episcopal Social Teaching

During the 1990s, national episcopal conferences and regional bishops' groups around the world made important contributions to Catholic

---

9. John Paul's concerns are reflected in a 1997 document issued by the Pontifical Council for Justice and Peace titled "Towards a Better Distribution of Land."

10. Cf. Greeley, *No Bigger Than Necessary*.

social teaching on sustainable development.[11] These documents reflected a sometimes slight, sometimes significant shift toward a just-sustainability approach to the eco-social question and systemic change. In the US, one remarkable example was the Appalachian bishops' pastoral message "At Home in the Web of Life" (1995).

The bishops' message merits attention for several reasons. First, "At Home" employed an inductive, contextualizing methodology that attended carefully to the mosaic of cultural traditions and indigenous voices within the region and incorporated insights from both the natural and social sciences in its reading of Appalachia's rich history. In section I, scriptural reflection on the integrity of creation was informed by both theocentric and evolutionary-ecological perspectives. The traditional *imago dei* doctrine and stewardship ethic was affirmed, yet it took on added new meaning when placed in a larger natural-historical and sociocultural context. "At Home" also expressed both a communitarian ethos and an arcadian imagination by creating a deep sense of place and community through its sensitive portrayal of a distinctive cultural heritage embedded within a complex, evolving natural landscape. The revelatory story of the mountains—its diverse flora and fauna and its many peoples—provides a narrative framework for understanding the human vocation in *this* place, *this* particular patch of God's good earth that its people call home. As (general) revelations of God, the mountains and their many wild creatures are appreciated for the ways in which they precede and exceed the human presence. "At Home" thus rendered more credible the Christian claim that the human relation to nature is anthropo-apical, not anthropocentric; humans inhabiting this special place are charged with moral oversight of a value-laden natural world. Catholic social teaching on the integrity of creation and stewardship is contextualized in a creative, responsible manner.[12]

Second, the bishops' message consistently adopted a green-communitarian perspective in its interpretation of the ways globalization and the emergence of world risk society affect Appalachia and other marginalized, exploited regions across the planet. The bishops' historical and structural analysis depicted the region as caught up in and impoverished by successive waves of aggressive, mobile capital. With the incoming wave, cheap land and labor are exploited to maximize profits, which are exported to power-centers

---

11. See, e.g., the various pastoral statements on sustainable development made by bishops from the United States, Guatemala, Australia, Northern Italy, Dominican Republic, and the Philippines collected in Christiansen and Grazer, *And God Saw*.

12. Cf. the Filipino bishops' pastoral letter on ecology, "What Is Happening to Our Beautiful Land" (January 29, 1988), http://www.cbcponline.net/documents/1980s/1988-ecology.html.

outside the region; with the outgoing wave, disinvestment leads to forced migration as workers and their families exit an exhausted landscape. In "At Home" the region is viewed as an exploited peripheral area within the global capitalist system, subject to chronic unemployment and harsh forms of environmental degradation (e.g., mountaintop removal mining)—both of which contribute to de-ruralization. Moreover, highly concentrated land ownership constitutes a structural constraint to community self-development efforts at county and state levels; at the time of writing (1995) over half the land surface was controlled by just 1 percent of the region's residents as well as absentee land-holders, corporations, and government agencies. Appalachia's working and middle classes thus remained dependent in many ways not only on local elites but also on powerful outside agencies, namely the corporate landowners and government bureaucracies that control much of the resource base.

In addition to its longstanding role as a source of timber and minerals, Appalachia has been targeted as a profitable dumping ground for industrial wastes. Other sources of profit for outside interests include superstores that tend to undermine locally owned and operated businesses as well as private prisons and gaming operations that typically create low-paying, dead-end jobs. Hence the recent incoming wave of investment promised more of the same: a further erosion of the region's natural and social ecologies. "At Home" made it clear that an ethical-political choice must be made between passive acceptance of this exploitative, crisis-ridden state of impoverishment and active commitment to the search for sustainable community.

Third, the bishops' commitment to a preferential option for the poor influenced not only their reading of the ongoing eco-social crisis in Appalachia, but also their ethical guidelines and strategic recommendations for a constructive alternative: building sustainable communities from the bottom up. The bishops' broad interpretation of subsidiarity emphasized the need to check the power of transnational corporations and implied clearly that their operations ought to be subject to democratic control at both local/regional and state/national levels so that new forms of homegrown social enterprise could compete on a fair playing field. "Just as political bureaucracies should not undermine local government," the bishops stated, "so business bureaucracies should not undermine local economics." Drawing attention to highly skewed patterns of land ownership, the bishops linked the just social use of property to the need for "responsible and legal land reform." And to concretize their belief that the common good is best served when "most property . . . is rooted in the local community," the bishops endorsed community land trusts.

Finally, "At Home" may be read as an exercise in prophetic-utopian imagination. Not content to criticize current trends, the bishops sought to discern disregarded possibilities hidden within the present situation and envision an "alternative future for the people and the land." Their "concrete utopia" (Bloch) involved a complete rethinking of the technology-economy-politics-ecology nexus around the concept of sustainable community: "The market needs to be rooted in the creative community of the local web of life. Its rooted place should not be eroded by governmental or corporate bureaucracies. Similarly the market needs to be guided by human dignity and by social and ecological community." In response to globalization's largely negative impact on the region, the bishops recommended a turn to cooperative, human-scale economics focused on building a strong "local social market" in which women's micro-enterprises and other civil-economic activities play an important role. This strategy aimed to democratize wealth and economic opportunity rather than extract resources and profits through exploitative practices. Toward this end, "At Home" held up several promising experiments in the fields of agro-ecology, sustainable forestry, cooperative land ownership, and "appropriate technology"—all of which reflect the vision and values of green communitarians.

A Weak Precautionary Stance

As noted in chapter 2, Beck and other risk-society theorists contend that industrial modernity generates a self-endangering progress, which in turn triggers political conflicts over how technological and environmental risks are defined and distributed. The promise and peril of emerging technologies—e.g., genomics, nanotechnology, robotics/AI—have raised the stakes in world risk society and generated a growing debate between techno-progressive advocates of a proactionary approach to exponential technology development and bio-conservative defenders of the precautionary principle, versions of which have become enshrined in environmental law with increasing frequency.

In 2004 the Pontifical Council for Justice and Peace released the *Compendium of the Social Doctrine of the Church*, which included a chapter titled "Safeguarding the Environment" that addressed the related issues of environmental protection and technological development within the broad context of the church's teaching on humanity's relationship to the created order in light of revelation and its reading of the signs of the times. Generally speaking, the chapter reflected a tilt in Catholic social teaching toward an eco-modernist approach to sustainable development, though some passages echoed green-communitarian concerns, e.g., the defense of the rights of indigenous peoples against the encroachments of "powerful agro-industrial interests or the powerful processes of assimilation and urbanization" (#471).

The church's continuing ambivalence toward techno-economic progress was reflected in the *Compendium*, which adopted what may be called a "weak precautionary stance" in response to a question introduced in chapter 2 that is central to the emerging proactionary-precautionary debate: In situations characterized by uncertainty and risk, what is the most responsible way to develop and assess technologies? To understand the church's position during this time, we must attend first to the way in which the *Compendium* frames the question through a broader discussion of the role of science and technology, the philosophical errors underlying the environmental crisis, and the need for a global ethic of responsible care.

Early on, "Safeguarding the Environment" reaffirms the church's positive, early-modern view of techno-economic progress: "[E]specially with the help of science and technology, man has extended his mastery over nearly the whole of nature and continues to do so" (#456) . . . as people who believe in God, who saw that nature which he had created was 'good', we rejoice in the technological and economic progress which people, using their intelligence, have managed to make" (#457). Such progress includes recent developments in bio-technology, though the *Compendium* quickly qualifies its general approval with the reminder that "it is necessary to maintain an attitude of prudence and attentively sift out the *nature, end and means* of the various forms of applied technology" (#458).

At this point, a telling distinction is drawn between morally legitimate forms of techno-scientific cooperation with the "design of God" in the "development" of living creatures and ethically unacceptable forms of genetic manipulation that do not honor "God's prior and original gift of the things that are." The former displays a respectful approach to nature's "prior God-given purpose, which man can indeed develop but must not betray," while with the latter "man sets himself up in place of God and thus ends up provoking a rebellion on the part of nature, which is more tyrannized than governed by him" (#460). Here and elsewhere, recent Catholic social teaching continues to rely on a metaphysical background theory, namely a Thomist metaphysics in which each and every creature has its place and purpose within a hierarchical order of being, a diverse array of beings that manifests the power, wisdom, goodness, and beauty of the Creator. In "The Ecological Crisis," for example, John Paul observed: "Theology, philosophy and science all speak of a harmonious universe, of a 'cosmos' endowed with its own integrity, its own internal, dynamic balance. *This order must be respected*" (#8). Given its reliance on a Thomist, teleological view of the order of creation—as well as its implicit agreement with an older model of ecology that emphasizes telic stability (homeostasis) rather than probabilistic change (stochasis)—the *Compendium* expresses a bio-conservative wariness toward genetic engineering and other forms of techno-scientific intervention in nature, an attitude in tension with its early-modern confidence in science and technology as drivers of industrial progress.[13]

The *Compendium* then draws a second, related distinction between scientific research and technological development that serves to promote integral human development while respecting the integrity of creation, on the one hand, and the serious errors of scientism and technocracy, on the other. With the latter, we arrive at the root cause of environmental problems, namely the rise in recent centuries of a materialistic worldview and technocratic paradigm in which nature is reduced to mere resource and its theo-ontological status as divine creation is denied in both economic practice and secular philosophy:

> *Nature appears as an instrument in the hands of man, a reality that he must constantly manipulate, especially by means of technology.* A reductionistic conception quickly spread, starting from the presupposition—which was seen to be erroneous—that an infinite quantity of energy and resources are available, that it is possible to renew them quickly, and that the negative effects of the exploitation of the natural order can be easily absorbed. This reductionistic conception views the natural world in mechanistic terms and sees development in terms of consumerism. Primacy is given to doing and having rather than to being, and this causes serious forms of human alienation (#462).

When a fundamental attitude of gratitude and appreciation toward creation—a loving gaze—is replaced by an aggressive interrogation of nature—a controlling gaze—openness to mystery is eclipsed by the will to domination and a "war" on nature ensues (cf. #487). The errors of modern anthropocentrism have in turn provoked their diametric opposite in ecocentric and biocentric philosophies that fail to respect the anthropo-apical status of human beings in the order of creation (#463). Having abandoned a theocentric perspective—one that relates and directs all beings to God as their source and goal—these modern, secular philosophies end up displacing humanity

---

13. From the perspective of contemporary ecological theory, church teaching's strong emphasis on the order, harmony, and alleged "balance" of nature seems outdated; see, e.g., Botkin, *Discordant Harmonies*. Departing from Odum's "ecology of order," Botkin and other ecologists now emphasize the role of periodic disturbances and stochastic processes in evolving ecosystems; cf. Deane-Drummond's critique in "Joining the Dance."

from its proper place/role as moral overseer and priest-celebrant of creation. Modern humanity is either estranged from nature-as-God's-good-creation due to anthropocentric hubris or robbed of its dignity as "crown of creation" by the false humility of radical environmentalists.

In light of the environmental crisis, the *Compendium* stresses the need for a global ethics of responsible care attentive to the fate of future generations. A shared destiny implies shared responsibility: "Care for the environment represents a challenge for all of humanity. It is a matter of a common and universal duty, that of respecting a common good" (#466). As the Amazon goes, so shall we and our progeny. A common responsibility extends to future generations and must be expressed not only through strong environmental standards and monitoring capacities established by governments, but also through an "effective change of mentality and lifestyle" throughout society (#468).

Within this broad context, then, the *Compendium* offers a fairly detailed and nuanced account of the precautionary principle and its use in environmental policy-making and technology assessment under conditions of uncertainty and risk. Emphasis is placed on the need for prudent judgment, with the precautionary principle viewed as a "guideline" rather than a decision rule for policymakers who must manage complex, evolving situations by making "temporary decisions" in an open, flexible manner based on a comparative weighing of the risks and benefits associated with each policy option, including no action (#469). Regarding the risks and benefits attending bio-technology, we read that "in the realm of technological-scientific interventions that have forceful and widespread impact on living organisms, with the possibility of significant long-term repercussions, it is unacceptable to act lightly or irresponsibly" (#473). Here the spirit of precaution clearly is present. However, no clear position is taken on the controversial burden-of-proof issue: In situations where significant, irreversible harms to human and environmental health may occur, should advocates of technological intervention bear the burden of proof for demonstrating safety up front (and at what level? zero risk?), as advocates of a "strong" precautionary principle argue, or is it acceptable to allow technological deployment to go forward in the (current) absence of clearly demonstrable harms, as defenders of standard risk-assessment methodology hold? The *Compendium* does not say, and we are left to wonder why, given the importance of the question in the ongoing debate.

A fair reading of the church's weak precautionary stance would point up the difficulty of taking decisions on public health and safety in situations of scientific uncertainty. To their credit, the *Compendium* authors rightly underscore the need for a transparent, flexible decision-making process informed by the best available science and governed by prudence rather than expediency or political influence (#479), and their global eco-justice perspective draws needed attention to just forms of technology transfer that avoid neocolonial exploitation and promote the "*development of a necessary scientific and technological autonomy*" within poor nations (#475). Nonetheless, the precautionary stance on technological and environmental risks adopted by the magisterium during this time was weak, and for two reasons.

First, as just noted, the *Compendium* takes no clear position on the crucial question concerning who bears responsibility for demonstrating that undue risks are not being taken with human lives and the natural life-support systems upon which they depend. One can reply, as many proactionaries do, that promising new technologies must be tried out in the field to discover their full, positive potential as well as their unintended (and most often unforeseeable) negative consequences. This point carries weight in a world sorely in need of technological innovations that may help to solve the problems of hunger, disease, and environmental degradation. Indeed, the *Compendium* takes note of technology's potential for just this reason. Yet in cases where a small yet real potential for significant, irreversible harm exists, does not prudence dictate a strong precautionary approach, e.g., one that invokes the maximin rule, which prioritizes avoidance of worst possible outcomes?

Second, the authors of "Safeguarding the Environment" fail to apply Catholic social tradition's critique of technocracy to modern risk-management theory and practice. A core tenet of technocracy is belief in human capacities to control nature and predict future outcomes; the authority granted to the technocrat derives from science-based expertise and claims of predictive control. While the "subversive science" (Shepard) of ecology as well as the new sciences of chaos and complexity have called this belief into question, the myth of controllability still functions as an unquestioned article of faith in a range of modern techno-scientific practices, among them the applied science of risk assessment.

Absent from the *Compendium*'s discussion of the precautionary principle is a critique of standard risk-assessment methods and processes, which are complicit in the "organized irresponsibility" (Beck) of science, law, business, and government—interlocking institutions driven by a capitalist growth imperative that hinges in no small part on technological innovation. Yet as noted in chapter 2, industrial and emerging technologies generate mega-hazards that put into question early-modern assumptions regarding the calculability of risks and reversibility of harms:

> Environmental degradation and hazards pose threats that differ from the calculable risks associated with car accidents, thefts

and house fires, in that they are neither just unintended consequences of rational actions, nor mere side effects, but endemic to the scientific innovations and economic practices characteristic of the industrial way of life: for hazards arising from the industrial way of life, the past gives no guidance for the future, provides no basis on which to calculate and quantify risk.[14]

Compelled by the growth imperative into a dangerous state of denial, risk-management practitioners within major institutions continue to live by the questionable creed that predictive control and reliable forecasting (within ranges of probability) is possible for any and all technological impacts. As a consequence, "modern forms of risk management . . . systematically underestimate unforeseen and improbable, but not therefore impossible, occurrences, as regards both their frequency and the extent of the damage they cause."[15] Put another way, adherence to the canons of risk assessment contributes to "black-swan blindness" (Taleb), i.e., a heavy discounting or complete denial of highly improbable, high-impact events that could and sometimes do prove catastrophic.

Lacking a critique of technocracy's dubious claims to controllability, and without a clear stand on burden-of-proof responsibility, the church's weak precautionary stance during this time was at risk of unwitting complicity with an irresponsible risk-management establishment that clings to faulty, early-modern assumptions. In *Laudato*, Francis departs from the *Compendium*'s position by arguing on eco-justice grounds for a strong version of the precautionary principle. As the next chapter shows, underlying this reversal is a critique of prevailing technocratic assumptions and practices, a critique informed by an ecological paradigm as well as Catholic social teaching dating back to Pius XII's warnings against the "technological spirit."

The Character of Tensions and Respectful, Generative Dialogue

In this chapter and the last, I traced certain tensions running through the church's evolving response to world risk society and its engagement with the sustainable-development and proactionary-precautionary debates. Such tensions reflect the realities of plural voices and conflicting camps within the church. It has always been thus, with conservatives and liberals

---

14. Adam, *Timescapes of Modernity*, 82.
15. Beck, *World at Risk*, 130.

advancing disparate and often diametrically opposed interpretations of how doctrine and tradition relate to modernity.[16] These internal tensions may be understood and evaluated as either constructive or debilitating in character. A constructive tension tends to generate further insights, promote ongoing dialogue, and open the way for a new synthesis or meaningful consensus, while a debilitating tension tends in the opposite direction and may yield any number of unhealthy outcomes. The former characterizes a learning church oriented to growth in understanding for the sake of more authentic Christian living, the latter a divided and dysfunctional church characterized by incoherent positions, one-sided readings of tradition, posturing, and stagnation.

In the church's response to the sustainable-development debate, an ongoing tension exists between support for a progressive, eco-modernizing approach that advocates major reforms aimed at transitioning to a high-efficiency, post-carbon economy and appeals primarily to elites within established institutions, on the one hand, and sympathy for a green-communitarian strategy of confronting social injustice through direct action and promoting alternative, community-based economic models with an eye toward displacing rather than reforming the institutions of corporate capitalism, on the other. Since the late 1980s, papal social teaching has tilted more toward eco-modernizers in its policy recommendations, yet also nodded respectfully to greens. Episcopal social teaching at the national/regional level, meanwhile, has shifted sometimes slightly, sometimes significantly toward a just-sustainability perspective and bottom-up strategy, though important elements of an eco-modernizing agenda are included as well.[17]

Social location goes some way toward explaining different prudential judgments on policy agendas and the use of one type of social theory rather than another. It is no accident, for instance, that "At Home in the Web of Life"—which adopts a green-communitarian outlook and draws heavily on a version of world systems theory—was issued by a group of bishops from a peripheral region with a long history of exploitation by outside interests. Nor is it surprising that John Paul II's writings reflected an eco-modern approach to sustainable development culturally underwritten by a Romantic nature aesthetic, given the resonance of this policy stance and attitude toward nature in Europe. Yet from a global church perspective, the tilts

---

16. Miller, "Catholic Social Teaching"; as Miller observes, "Catholic Social Teaching arises from a long, dynamic, and contentious tradition of Catholic social thinkers."

17. A good example of a balanced appropriation of both policy narratives is the Pacific Northwest bishops' pastoral letter, "The Columbia River Watershed: Caring for Creation and the Common Good" (2000), www.thewscc.org/uploads/3/4/9/4/34945816/colrvr-e.pdf.

and shifts toward one policy narrative or the other are less important than the ongoing affirmation of both perspectives in the context of respectful, generative dialogue. Rather than signaling confusion or indecision among church leaders, this dual affirmation reflects a constructive tension oriented to growth in understanding. The dialogical unfolding of this constructive tension over the past quarter-century has made it possible for recent Catholic social teaching, and here I would point to *Laudato* in particular as exemplary, to incorporate new insights and introduce fresh perspectives—notably an ecological paradigm—into its evolving social-ethical vision of integral human development.

In recent decades a learning church has sustained the development of its social teaching by engaging in respectful, generative dialogue with diverse perspectives and by drawing on different sources of insight—Scripture, tradition, and reflection on experience—as well as the metaphorical-affective power of three forms of imagination: communitarian, prophetic-utopian, and arcadian. Going forward my wager is that, as in the past, especially fruitful insights will arise from the pioneering efforts of socially conscious Catholics and other Christians who take the lead by inventing new forms of eco-social praxis and exercising moral leadership in a variety of settings, both privileged and impoverished.[18]

---

18. Here I would hold up as exemplary the recent eco-social Catholic leadership exercised by many women religious in the US and elsewhere; see Taylor, *Green Sisters*.

5

The Logic of Gift and Integral Ecology

Let's begin with a tale of two generations. In recent years Kelly Services has used its digital platform to place over 500,000 blue-, pink-, and white-collar workers in temporary roles. The temp-services giant fills over two million US classrooms each year with substitute teachers. CEO Carl Camden says the company serves "as the HR department for the contingent workforces of over 90 percent of the corporations in the Fortune 500."[1] On his many recruiting visits to college campuses, Camden preaches liberation from wage slavery through a more flexible, entrepreneurial approach to work and career. His gospel of free agency resonates with many Millennials, while others hanker for more secure futures.

Camden's son Andrew has started his own company, Detroit Drones, while holding down a six-figure engineering job at General Motors. Andrew and a friend use a 3D printer to design drone components. What distinguishes father from son is the latter's commitment to open-source software. While his father cannot fathom why anyone would give away the company's IP for free, Andrew gets his biggest kick from positive feedback provided by users of his design files, which can be accessed online at no charge. How might Detroit Drones monetize its offerings or get bought out by a digital behemoth? These are not burning questions. Value created and shared brings satisfaction. The world doesn't need a General Drones.

Perhaps Andrew's passion is just a hobby masquerading as a startup. Or maybe his company and thousands of other new enterprises are pioneers in an ongoing shift toward new forms of social production that will generate and share economic value—real goods and services—freely with all comers rather than extracting value from not-so-free workers and consumers.

---

1. Stern, *Raising the Floor*, 80.

Recent decades have seen the growth of a civil market economy characterized broadly by an ethos of sharing, abundance mindset, open-source innovation, distributed networks, self-governing structures, and cooperative provision.[2] Benefit corporations, worker-owned cooperatives, health care sharing ministries, and other forms of social enterprise pursue diverse missions, yet they generally share a desire to address needs unmet by markets and governments, democratize wealth and decision-making power, create community, and operate in an environmentally sensitive manner. While it remains in the shadow of the formal economy, the civil market economy's growth potential ought not to be underestimated. In *The Wealth of Networks*, Yochai Benkler demonstrates how the digital revolution has created favorable conditions for economically viable social production. As an alternative means for generating economic value, social production may complement profit-driven enterprise and at other times compete successfully against private, proprietary models.[3]

Meanwhile, the Great Acceleration continues apace. World GDP over the past half-century has grown from $23 trillion in 1975 to $75 trillion in 2016. According to International Monetary Fund projections, the world economy in 2050 will be 3.5 to 5 times larger than in 2010.[4] And although economic globalization has led to lower poverty levels and improvements in health care and education in a number of countries, global development remains starkly uneven with widening socioeconomic inequality both within and between nations. Islands of prosperity float uneasily amidst oceans of poverty.

As we noted in chapter 2, mounting environmental stress and growing inequality have lent more urgency to calls for new approaches to sustainable development. World leaders met in September 2015 under UN auspices to sign a joint, legally non-binding declaration that commits signatory states to working collaboratively on meeting seventeen Sustainable Development Goals by 2030. In policy circles, ecological modernizers in the EU and elsewhere lobby for a rapid transition to a clean-tech, post-carbon economy, often citing scientific research based on the planetary boundaries approach, which "aims to define a safe operating space for human societies to develop and thrive, based on our evolving understanding of the functioning and

---

2. As Zamagni suggests, late-medieval Europe's nascent civil market economy, before its slow strangulation by aristocratic and nation-state building elites, offers a relevant historical precedent for today's small but growing civil market economy; see Zamagni, "Civil Economy." A similar argument is developed by Rushkoff in *Throwing Rocks* and by Milbank and Pabst in *Politics of Virtue*.

3. Benkler, *Wealth of Networks*.

4. Gilding, *Great Disruption*, 50–51.

resilience of the Earth system."⁵ The latest research indicates that one of two core boundaries—biosphere integrity—has moved into the high-risk danger zone along with biogeochemical flows, while the other core boundary—climate change—and land-use changes are now in the zone of uncertainty with increasing risk of negative impacts.⁶

Recent Catholic social teaching has responded to these developments by calling (again) for wholesale changes in mindsets, lifestyles, social practices, and institutional policies. Papal and episcopal social teaching over the past quarter-century, I have argued, has maintained a constructive tension between eco-modern and green viewpoints. The ongoing dialogue within the world church has generated new insights and introduced fresh perspectives that find expression in the social encyclicals of Benedict XVI and Francis.

Benedict's articulation of the logic of gift provides a powerful theological-ethical rationale for social innovation and expansion of a growing civil market economy. Such innovation and expansion promises to play an increasingly important role in any serious effort to create a diverse, hybrid economy and socially inclusive post-jobs society. Together with Benedict's vision of an "economy of communion," the welcome introduction of an ecological paradigm into Catholic social teaching by Francis opens the way for a careful consideration of strategic alternatives to business as usual (i.e., a crisis-ridden continuation of neoliberal globalization). In particular, the idea and ideal of integral ecology articulated by Francis can inform and inspire Christian praxis and moral leadership aimed at building sustainable communities, reinventing institutions, and resynchronizing natural and social ecologies in many different contexts, privileged as well as impoverished, across the planet.

Toward an Economy of Communion

Released in 2009, Pope Benedict's social encyclical *Caritas in veritate* focused on the complex, evolving relationships between market, state, and civil society in a global economy still mired in the Great Recession. Benedict made a distinctive theological contribution to Catholic social teaching by elucidating the ethical-practical implications of what he called the "principle of gratuitousness and the logic of gift." The gift of God's superabundant grace heals and empowers human persons (who already are gifted "originally" with a pro-social, giving nature), enabling them to move

5. Steffen et al., "Planetary Boundaries."
6. Ibid.

beyond relationships guided solely by legal-formal norms of strict reciprocity—the logic of exchange—to relationships based on generous sharing—the logic of gift—that create bonds of trust and enhance community well-being. The former is transactional in character, benefiting individuals, while the latter is transformational in its capacity to generate friendships and strengthen the common good. Here Benedict applies to economic life a classic Christian claim: love requires justice as its minimal standard, indeed demands it; yet love also transcends justice for the sake of reconciliation and communion. Hence the aspiration to build an economy of communion that aims not at utility maximization for many individuals, but rather at the co-creation through meaningful work of a sustainable abundance enjoyed in fair measure by all.

The principle of gratuitousness arises from and is confirmed by the human cultural experience of giving and receiving gifts, a cross-cultural phenomenon constitutive of community.[7] As recipients of a natural-supernatural grace that evokes gratitude, joy, and a sense of responsibility, Christians are called to live by the go-and-do-likewise logic of gift in both the interpersonal and social spheres. For Benedict, the "dynamic of charity received and given is what gives rise to the Church's social teaching, which is *caritas in veritate in re sociali*: the proclamation of the truth of Christ's love in society" (CV #5). Thus, with the continuing challenges of global development in view, *Caritas* calls for new forms of economic practice that reflect the logic of gift:

> The great challenge before us, accentuated by the problems of development in this global era and made even more urgent by the economic and financial crisis, is to demonstrate, in thinking and behaviour, not only that traditional principles of social ethics like transparency, honesty and responsibility cannot be ignored or attenuated, but also that in *commercial relationships* the *principle of gratuitousness* and the logic of gift as an expression of fraternity can and must *find their place within normal economic activity*. (CV #36)

This and similar statements apply most readily to an emerging civil market economy in which the principle of gratuitousness finds concrete expression in the creation of both real wealth and high-trust community. *Caritas* follows tradition in noting that, beyond the market logic of exchange as regulated by commutative justice, governments play a legitimate role as regulators of distributive and legal (or contributive) justice. The new accent comes with a theological-ethical appeal to move political-economic

7. Mauss, *The Gift*.

structures and institutions beyond standard business logic and traditional canons of justice by making room for a truly humanized economics based on the principle of gratuitousness: "Space also needs to be created within the market for economic activity carried out by subjects who freely choose to act according to principles other than those of pure profit, without sacrificing the production of economic value in the process" (CV #37). Toward this end, governments should establish a legal-regulatory framework for a civil market economy that puts social enterprises, cooperatives, not-for-profits, and other civil-society organizations on an equal footing with for-profit businesses:

> What is needed . . . is a market that permits the free operation, in conditions of equal opportunity, of enterprises in pursuit of different institutional ends. Alongside profit-oriented private enterprise and the various types of public enterprise, there must be room for commercial entities based on mutualist principles and pursuing social ends to take root and express themselves. It is from their reciprocal encounter in the marketplace that one may expect hybrid forms of commercial behaviour to emerge, and hence an attentiveness to ways of *civilizing the economy.* (CV #38)

In this passage we glimpse the possibility of a diverse, hybrid economy no longer dominated by either state or corporate monopolies. The passage of state laws that allow benefit (or simply "B") corporations to pursue a triple bottom line—rather than being legally required to advance first and foremost the financial interests of shareholders—is a clear step toward what the pope has in mind. As of 2016, B corporation laws exist in thirty-one U.S. states and are being proposed in others.

The growth of a civil market economy, Benedict insists, should not be limited to advanced-industrial countries, for there are limits to what (capitalist) markets and states can do to eradicate mass poverty and marginalization in the global South. Civil-economic initiatives offer a promising alternative: "In order to defeat underdevelopment, action is required not only on improving exchange-based transactions and implanting public welfare structures, but above all on gradually increasing openness, in a world context, to forms of economic activity marked by quotas of gratuitousness and communion" (CV #39).

Caritas is not without its critics. Benedict has been taken to task for overemphasizing personal responsibility while underplaying the structural character of inequality and need for collective action. The pope's implied theory of social change leans heavily, once again, on consensus-building and reforms made by elites presumably responsive to reason and moral appeals. And methodologically the encyclical takes a deductive, top-down approach that fails to include the voices and perspectives of local churches. Nonetheless, Benedict's articulation of the logic of gift in economic life is a welcome restatement of Catholic social wisdom, one that reaffirms the church's long-standing support for cooperative economic activity and holds up a hopeful sign of the times. Moreover, expanding the civil market economy accords with a growing body of research in evolutionary theory, anthropology, positive psychology, and behavioral economics that highlights the role of cooperation, social norms, non-material incentives, and giving behaviors.[8]

In sum, the principle of gratuitousness provides theological-ethical grounds for social innovation and further growth of the civil market economy.[9] This opening to and support for an alternative pathway toward the creation and equitable sharing of wealth within communities is an important point of departure for Christian praxis now and in the future. In chapter 7, I situate a growing civil market economy within a broader strategic vision for social transformation.

---

8. See, e.g., Bowles and Gintis, *Cooperative Species*; Ariely, *Predictably Irrational*, chap. 4; and Grant, *Give and Take*.

9. Cf. the significant postliberal proposal for (re)constructing a full-blown civil market economy in Great Britain (and beyond) elaborated by Milbank and Pabst in *Politics of Virtue*, chap. 4. The authors draw heavily on Catholic social thought, both medieval and modern, in calling for a recommitment to the "priority of the social" in both economics and politics, i.e., to re-embedding both economic practice and political life in (primarily local) communities founded upon and sustained by an ethos of generous sharing and a shared commitment to excellence in all the arts (i.e., the agricultural, industrial, and political arts as well as high and low cultural activity). They also argue convincingly for a renewal of culture as formation, arguing that schools and other cultural institutions must perform an educative function aimed at achieving what might be termed a "deep cultural literacy" that forms the virtuous character of both the community and its members. While in later chapters my priorities and proposals for social action, policy reform, and cultural renewal differ somewhat from that of Milbank and Pabst—e.g., I see more potential in emerging technologies, defend a universal basic income, and so welcome the coming of a post-jobs society in which comparatively more time is devoted to activities other than wage labor, whereas they seem to want a civil market economy that deploys "appropriate technologies" and achieves full employment—I am sympathetic to their project. Substantive agreement exists on matters of fundamental importance, not least of which is: (1) the grounding of both economics and politics in a more adequate view of human sociality; (2) the desirability of a wider distribution of wealth and power in society under the principle of subsidiarity; and (3) the compelling need for a postliberal cultural renewal focused on character formation through practices, a communal ethos of generosity and solidarity, and shared commitment to excellence in all the arts.

The Idea and Ideal of Integral Ecology

Laudato si' (2015) is the first papal encyclical devoted primarily to the eco-social question and global environmental crises. While *Laudato* carries forward previous Catholic social teaching on responsible stewardship, the common good, and other core elements of the tradition, it also breaks new ground in both method and substance. Methodologically, Francis takes an ecumenical, collegial, and dialogical approach, e.g., by including Patriarch Bartholomew's prophetic voice and many bishops' statements on ecology and justice from around the world. *Laudato* deftly arranges these voices to create a powerful eco-justice chorus and amplify the encyclical's call to ecological conversion and a bold cultural revolution. Francis also adapts the familiar see-judge-act method to his own purposes by drawing extensively on findings from ecology, conservation biology, and related scientific disciplines in his discussions of pollution, climate change, fresh water security, loss of biodiversity, and other environmental problems.

Yet there is more to Francis's engagement with the scientific community than a summary of the current consensus on climate change, Holocene extinction, deforestation, and other matters of grave concern to global society. The background theory evident throughout *Laudato* is an ecological paradigm, which today informs ongoing research in sustainability science. Adoption of this perspective leads Francis to emphasize, among other things, the relational dimension of existence:

> It cannot be emphasized enough how everything is interconnected. Time and space are not independent of one another, and not even atoms or subatomic particles can be considered in isolation. Just as the different aspects of the planet—physical, chemical and biological—are interrelated, so too living species are part of a network which we will never fully explore and understand. A good part of our genetic code is shared by many living beings. It follows that the fragmentation of knowledge and the isolation of bits of information can actually become a form of ignorance, unless they are integrated into a broader vision of reality. (LS #138)

Aiming for a broader vision of reality, sustainability science is an emerging interdisciplinary research program informed by an ecological paradigm that challenges the reductive, mechanistic assumptions of modern scientific methodology. Is and ought, fact and value, theory and practice, analysis and advocacy—these pairs are viewed as complementary and even interfused in sustainability science rather than being taken as markers separating the sciences from the humanities and policy-making.[10]

Viewed philosophically, sustainability science presumes a social, process-oriented ontology from which normative commitments follow. Human persons are always already constituted by a network of relationships within communities, co-developing with other beings-in-process and co-inhabiting landscapes that are themselves continually shaped and reshaped by interacting geological, bio-chemical, atmospheric, and sociocultural processes operating at multiple spatio-temporal scales. Our epidermis and ego can be viewed as semi-permeable boundaries through which flow molecules, nutrients, energies, affects, information, and insights that connect us in myriad ways to countless other entities—all of which are alive and responsive in some degree of intensity and richness. Human bodies are micro-biomes living in symbiosis with countless other life forms. Human understanding takes manifold form, developing as much through embodied, empathic, and adaptive responses to an ever-changing environment as through Cartesian detachment, objectification, and control. Here the master metaphor for nature shifts from a vast machine to an array of networks or fields through which energies and information continually flow and within which self-organizing entities and systems emerge as conditions allow. Earth is thus seen as a complex, interconnected, and fragile web of life with limits to its resilience. It follows that humanity transgresses planetary boundaries at its peril.

The ecological paradigm informing *Laudato* finds expression in the idea and ideal of integral ecology, an expansive concept that turns on the recognition and appreciation of interconnectedness at all levels of existence. Importantly, integral ecology implies normative judgments about the nature and quality of relationships existing within society and between natural and social systems. In keeping with both Catholic social principles and operative norms within sustainability science, Francis points up the need for cooperation, appropriate scale and tempo, and distributed power as counterpoints to unrestrained competition, inordinate growth and acceleration, and overly concentrated power. In human affairs, individual striving and competition are kept healthy within the sustaining and constraining context of mutual support and social inclusion; good policies and practices strike

---

10. While Francis does not use the term *sustainability science* explicitly, *Laudato* is informed deeply by this new, problem-based approach to the study of earth systems. The drafting process for the encyclical included extensive consultations with representatives from the natural and social sciences. In 2014 the Pontifical Academy of Sciences and the Pontifical Academy of Social Sciences held a joint workshop at which leading practitioners reported on the current state and major findings of sustainability science.

the right balance between competition and cooperation, individual freedom and collective security. Similarly, the dynamism and creativity associated with urbanization, technological innovation, and economic growth remains vibrant when balanced by attention to optimal scale and respect for the intrinsic temporalities of culture and ecology. Human communities face the challenge of finding a situated fitness within ecosystems and the biosphere by designing with nature and calibrating resource use to nature's productive, assimilative, and regenerative capacities through practices of conservation and adaptive ecosystem management.[11] And wherever economic and political power becomes too concentrated, democratic forces need to demand and struggle for a wider distribution of benefits, opportunities, and decision-making authority in society in keeping with the principle of subsidiarity.

In his discussion of environmental problems, Francis takes a synthetic, evaluative approach that renders various connections between natural processes and conditions, economic practices, social trends, political decisions, and cultural attitudes more visible and comprehensible. This approach reflects a moral epistemology, i.e., a way of perceiving, interpreting, and evaluating one's surroundings and situation in the round, as it were, and with an embodied, empathic response to the needs and vulnerabilities of others (including other species). Early on, for example, Francis invites his audience to consider what is going on in order to "become painfully aware, to dare to turn what is happening to the world into our own personal suffering and thus to discover what each of us can do about it" (LS #19). Later in the text, the pope reminds readers of their capacity for embodied cognition and empathy: "Thanks to our bodies, God has joined us so closely to the world around us that we can feel the desertification of the soil almost as a physical ailment, and the extinction of a species as a painful disfigurement" (LS #89). Integral ecology thus implies an ethos and sensibility—a profoundly affiliative way of being with others in the world—quite different from the dominant liberal ethos of individual advancement and technological mastery.

While the idea of integral ecology is in continuity with the Catholic social tradition's normative vision of integral human development, Francis moves beyond his predecessors by situating the integral scale of (human) values within a value-laden natural world and decidedly shifting the emphasis from domination to dependence in his view of the humanity-nature relationship. Indeed, *Laudato* is remarkable for its explicit affirmation of what environmental philosopher Holmes Rolston calls the "autonomous

---

11. On how practices of adaptive ecosystem management require openness to cultural criticism and transformation of the kind called for in *Laudato*, see Jenkins, *Future of Ethics*, chap. 4.

intrinsic value" of individual organisms, species, and ecosystems.[12] According to the pope,

> We take these systems into account not only to determine how best to use them, but also because they have an intrinsic value independent of their usefulness. Each organism, as a creature of God, is good and admirable in itself; the same is true of the harmonious ensemble of organisms existing in a defined space and functioning as a system. Although we are often not aware of it, we depend on these larger systems for our own existence. (LS #140)

As noted above, sustainability science does not permit sharp lines, methodological or ontological, to be drawn between is and ought, fact and value. One's epistemic stance is that of a sensitive naturalist for whom, as Rolston puts it, "an 'ought' is not so much derived from an 'is' as discovered simultaneously with it."[13] *Laudato* stresses human duties to protect species and ecosystems based not on enlightened self-interest, but rather on a theologically-grounded affirmation of intrinsic values in the natural world (LS #69).

In the policy arena, integral ecology provides a broad framework for further dialogue and collaboration between progressive ecological modernizers and green communitarians on alternative approaches to sustainable development, e.g., through biomimicry and adoption of closed-loop production processes (LS #22), a rapid transition to renewable energy systems (LS #26), and promotion of sustainable agriculture (LS #180). Francis supports a pluralist approach to systemic change, arguing for the complementarity of top-down and bottom-up strategies. Formation of a "true world political authority" capable of establishing and maintaining strong regulatory regimes must respect the principle of subsidiarity and cannot succeed without democratic renewal and continual pressure on elites from civil-society forces.

Laudato reflects the constructive tension between eco-modernizing and green-communitarian approaches in recent Catholic social teaching. Both the pollution-prevention-pays logic of the former (LS #190–92) and global eco-justice demands of the latter are voiced. For example, Francis echoes recent statements made by the World Council of Churches that point up the "ecological debt" owed by the global North, which continues to reap disproportionate benefits from resource exploitation in the global South and, with regard to climate change, has used up a large share of the

---

12. Rolston, *Environmental Ethics*.

13. Rolston, "Ecological Ethic?," 101. For a defense of Rolston's axiological and epistemological claims, see Nunez, "Intrinsic Value of Nature."

atmospheric commons through intensive industrial development over the past two centuries. Accumulating ecological debt places upon developed nations a greater (or differentiated) responsibility for transitioning to less consumptive domestic economies while also offering substantial aid to developing countries for poverty elimination as well as their own transition to post-carbon, clean-tech economies.

These daunting transitions cannot be pursued, however, in isolation from the concerns of social, cultural, and human ecology. For Francis, social ecology points to levels of social trust, the health of governing institutions, and respect for law and morality within society. When a political culture falls captive to a narrow ideology or becomes polarized, government cannot regulate community affairs, rein in the powerful, and combat corruption. The poor and the earth suffer the consequences. The pope also insists that efforts to address socioeconomic inequality and environmental degradation must take into account the cultural meaning of ecology at a time when globalization now threatens cultural diversity as much as biodiversity. Much more is at stake than preservation of historic cultural sites and artifacts. Development projects need to respect local, living cultures and not impose alien, consumerist conceptions of the good life. In particular, indigenous communities should participate in planning as "principal dialogue partners, especially when large projects affecting their land are proposed" (LS #146). In the context of growing pressure to exploit arable lands, forests, and mineral deposits as quickly and cheaply as possible, *Laudato* reaffirms the church's strong support for indigenous land rights.

Finally, *Laudato* addresses several aspects of the "ecology of daily life." In line with Paul VI and John Paul II, both of whom spoke to issues of human ecology, Francis draws attention to the impact of the built environment and city landscapes on human well-being. While basic needs such as clean water and affordable housing take priority, still we desire to dwell in beauty and inhabit places that foster a distinctive community character. The quality of life depends on architectural designs and urban planning attuned to the vernacular, to civic flows and social intercourse, to an abiding sense of place.

To sum up, the idea and ideal of integral ecology provides a high-level, synthetic vision for integral human development and sustainable community. Francis calls for dialogue, collaboration, and empathic understanding in the interdisciplinary practice of a sustainability science that can develop integrated solutions to local-global problems based on an ecological paradigm and an authentic humanism open to the transcendent dimension. Importantly, the envisioned paradigm shift to an "ecological culture" must confront the root causes of the environmental crisis, which for Francis are traced to the "dominant technocratic paradigm" that has shaped the modern

West. What the "dominant technocratic paradigm" refers to concretely is not entirely clear, however, since the pope's discussion of modern technology and its ills often proceeds at a high level of generality.

Clarifying and Concretizing the Critique of Technocracy

Francis's critique of technocracy focuses on issues of power and the problematic relationship between modern technology and the good life. Drawing on the work of Albert Borgmann, we can clarify the specific character of our predicament and identify points of leverage for resisting the rule of technocracy. I begin with a reading of Borgmann's philosophy of technology and then show how it sheds light on *Laudato* by rendering more concrete what is somewhat abstract in the papal document.

Borgmann works out his own position on what he calls the "device paradigm" (or "pattern of technology") in relation to two prominent viewpoints, substantive and instrumental. The radical, substantive view of technology finds its most forceful articulation in Ellul, who sees *technique* as "the autonomous and irresistible power that enslaves everything from science to art, from labor to leisure, from economics to politics."[14] Nothing escapes the logic of efficiency inherent in technique. The imperative of greater efficiency, power, and speed drives modern technological development regardless of stated ends; nature (including human nature) is subordinated systematically to technique, or so Ellul argues. By contrast, the mainstream view holds that technologies are simply value-neutral tools. An instrumental approach to technology is "congenial to that liberal democratic tradition which holds that it is the task of the state to provide means for the good life but wants to leave to private efforts the establishment and pursuit of ultimate values."[15] Government ensures security and promotes prosperity in part through scientific research and the development of new technologies (e.g., through the work of DARPA). (Presumably a utilitarian calculus guides the decision-making and production process, though in recent decades US public funding has flowed more toward corporate-sponsored research projects.)

If the substantive viewpoint leaves the distinct character of modern technology obscure and appears too deterministic, the instrumentalist approach suffers from superficiality and the mistaken notion that with the right values in place, society can harness technology for the good. Neither irresistible power nor mere instrument, Borgmann holds that technology is characterized more accurately as a pervasive pattern shaping both

14. Borgmann, *Technology*, 9.
15. Ibid., 10. Cf. the critique of liberalism in Milbank and Pabst, *Politics of Virtue*.

technological production and everyday life in modern society. Technological devices provide us with quick, ubiquitous, safe, and easy access to commodities of all kinds. Flip the switch, turn the key (or dial), press the button, swipe the card—it's there, guaranteed. With its machinery concealed, the device at hand reliably delivers the goods (or services) and thereby disburdens us from arduous self-provision. No more chopping wood to stay warm, no more long walks to and from the watering hole, no more dark and screenless nights.

According to Borgmann, the origins of the device paradigm can be traced to the Enlightenment, a movement of intellectual and cultural liberation that took as its practical corollary freedom "from disease, hunger, and toil, and of enriching life with learning, art, and athletics."[16] Materially, the Enlightenment meant lightening the load of manual labor so that "higher pleasures" (Mill) might be pursued; technology promised liberation for the sake of self-development. Relief from the drudgery of mundane, physical work and release from the provincialism of village life is a real gain, Borgmann readily concedes. Yet the promise of technology first articulated by Bacon, Descartes, and Campanella eventually gave way to the irony of mindless labor and commodified leisure for the many. In the twentieth century, "things in their depth yield to shallow commodities, and our once profound and manifold engagement with the world is reduced to narrow points of contact in labor and consumption."[17] While expanding certain possibilities for self-development (no film, no Hitchcock; no airplane, no Earhart; no Twin Towers, no Petit toeing a wire between them), modern technology has in the main created a taken-for-granted pattern and way of life marked by disengagement, distraction, frivolous comfort, and ennui.

On the side of labor, technological innovation and automation tends systematically toward de-skilling and/or eliminating workers: "Wherever there is a traditional area of skillful work, it is disassembled, reconstructed, largely turned over to machines, and artisans are replaced by unskilled laborers."[18] Many jobs demand little from workers, who in turn take little pride in what they produce and largely view their labor as merely a means to consumption. Borgmann's point is confirmed by Gallup surveys of the US workforce, which report low levels of employee engagement year after year.

On the side of consumption, the dubious doctrine of consumer sovereignty hides a widespread complicity with the device paradigm and its debilitating effects. Disburdened and enthralled by technology, few bother

16. Ibid., 36.
17. Ibid., 77.
18. Ibid., 118.

with backyard gardens, home-cooked meals, and a familial culture of the table; few take up an instrument or take to the floor for a dance after the meal; and few take the time to walk about and get lost in wonder on starry nights. Exchanging wilderness for Wii, the privileged pass up the opportunity to pursue worthier and more fulfilling lives. The pattern prevails by consent despite the fact that "avowed happiness appears to decline as technological affluence rises."[19] Hence making advanced technologies more widely available and directing them to better ends is not the most fundamental challenge we face.

If entry into the digital age has appeared to renew the promise of technology once more, it also has expanded dramatically the reach of the device paradigm by providing greater numbers entry into the global market and multiplying options for "amusing and informing ourselves to death," as media critic Neil Postman puts it.[20] Borgmann anticipates that "in the final extension of the paradigm, the globe itself must be seen and treated as one technological device."[21] Here the harbingers include Google Earth, the fast-developing internet of things, and rapid spread of sophisticated surveillance systems around the world. Moreover, as discussed in chapter 2, we appear to have entered a phase of hyperacceleration in which the pressure to make good on the (false) promise of a "full life"—i.e., one that maximizes the number and variety of "satisfying experiences" consumed by the psychological subject—is intensified, rendering the privileged more dependent than ever on technological devices and the corporations that deliver them.

In ways subtle and overt, the pattern of technology provides cover, as it were, for the dark side of capitalism. A moral condemnation of the system's stoking of greed, sanctioning of social injustice, and despoliation of nature does little good. For most people, evidence of social inequities and ecological damage does not dislodge the perceived normalcy and self-evident benefits

---

19. Ibid., 124. The empirical validity of the Easterlin paradox (i.e., the claim that rising affluence in technological society leads to less happiness, as measured subjectively through self-reporting), which Borgmann takes as well-established, has in recent decades been questioned. On the debate over happiness and affluence, see Brockmann and Delhey, *Human Happiness*. That said, Borgmann's main thesis rests finally on philosophical and theological grounds: the focal concerns commanding our attention are "other and greater than ourselves," i.e., profound realities with which we seek a deeper, more intimate encounter so as to find true happiness. Lurking beneath Borgmann's appeal to focal concerns is Augustine's *ordo amoris*.

20. Lance Strate, *Amazing Ourselves*, 78. Postman's media-ecological critique of TV and computers/Internet complements Borgmann's analysis; both defend a humanist tradition and embattled culture of literacy against the distracting immediacy and superficiality of contemporary digital culture.

21. Borgmann, *Technology*, 146.

of maintaining a ladder of affluence upon which all may climb "upward"—eventually. Moreover, modern moral theory is ill-equipped to handle the primary ethical question raised by the device paradigm: what is a good life? Granted, questions of utility and of rights and duties *within* the pattern of technology have validity, where for example matters of safety and equitable access arise; the democratic machinery is geared to addressing those issues, even if it often does so poorly. But the deeper and more demanding challenge is whether and how a reform of the device paradigm itself is possible, since the predicament of the privileged is precisely the irony of technology, i.e., the terrible yet hidden cost of disburdenment and unworthy condition of disengagement, distraction, and debility so many find themselves in.

Genuine reform begins by exposing the pattern of technology and placing the question of the good life front and center: "A morally trenchant conversation about the good life requires . . . that the pattern of our actions which is disguised and diffracted in the prevailing moral discourse is itself made the moral issue."[22] How to conceive of and pursue truly good lives is the most basic challenge we face. We face (or avoid) the existential task of making a "strong evaluation" (Taylor) regarding what counts as truly significant and of ultimate concern for us. Borgmann holds that flourishing is found in a life patterned by *engagement* in meaningful work, deep play, and fruitful leisure rather than mindless labor and distracting consumption.

In search of the good life, it will not do to dismantle or abandon modern technology, Borgmann argues. A push for more direct, hands-on engagement with technology by promoting the arts of maintenance also has its limits, since advanced "technology systematically withdraws devices or their machinery from our competence and care by making technological objects maintenance free, discardable, or forbiddingly intricate."[23] Rather, the way forward is through purposeful engagement with *focal things and practices*, which throws the pervasive pattern into sharp relief and thereby enables us to put technology in its place. To illustrate simply, the daily jog or walk with a friend through the neighborhood discloses a world, weaving one into the fabric of a community and deepening one's attachments and sense of place. A solitary treadmill workout in one's basement pales by comparison, even if the physiological effects are similar. The former involves physical and social engagement with focal things that shape sensibility and character as well as waistline. The latter yields a cardiovascular benefit yet amounts to a device-aided disengagement from the wider world, which in turn prompts the need for an iPod or some other device that delivers distraction.

22. Ibid., 173.
23. Ibid., 161.

Leverage points for a reform of technology appear when focal things and the worlds they constitute come into consciousness, perhaps because they are threatened or simply because their presence commands our attention.[24] Focal things such as gardens, parks, city squares, horse farms (or simply the horses), wild rivers (or simply the fish), concert halls (or simply the instruments), ball fields (or simply the gloves and bats), and worship centers provide meaningful contexts with varying degrees of complexity and richness. "Commanding presence, continuity with the world and centering power are the signs of focal things."[25] Focal things host memorable events and disclose a world of depth that simultaneously grounds and orients us, summoning our powers and connecting us to others through shared practices and disciplines that serve to cultivate virtues and extend human excellence; they are privileged sites for encounters with ultimate concerns that evoke wonder, awe, joy, and gratitude. Backpacking, woodworking, child-raising, tennis playing, painting, grassroots organizing, worshiping—these and many other focal practices "counter technology in its patterned pervasiveness and . . . guard focal things in their depth and integrity."[26] What is more, resolute commitment to a community of focal practice and its tradition offers a sense of direction and purpose; one becomes part of and contributes to a rich, ongoing narrative—the community's living history—while forming lasting friendships grounded in mutual support, shared ideals, and the pursuit of worthy ends. Through sustained engagement one undergoes re-habituation outside the device paradigm's mode of commodification.[27] In chapter 10, I explore prospects for (re)engaging with

24. Borgmann's neo-Heideggerian approach is evident in the following passage: "[T]he simple things will come to the fore only as the rule of technology is raised from its anonymity, is disclosed as the orthodoxy that heretofore has been taken for granted and allowed to remain invisible. As long as we overlook the tightly patterned character of technology and believe that we live in a world of endlessly open and rich opportunities, as long as we ignore the definite ways in which we, acting technologically, have worked out the promise of technology and remain vaguely enthralled by that promise, so long simple things will seem burdensome, confining, and drab. But if we realize the central vacuity of advanced technology, that emptiness can become the opening for focal things" (199).

25. Borgmann, *Postmodern Divide*, 119.

26. Ibid., 209–10.

27. On the challenges faced by persons of faith and religious communities in resisting the structures and habits of commodification in late capitalism, see Miller, *Consuming Religion*. Miller analyzes how ordinary life within a single-family house in the suburbs (a taken-for-granted structural reality), which typically includes daily doses of corporate culture customized to age cohorts, massively conditions the broad, self-identifying middle class into habits of consumption that erode or distort capacities to engage deeply with focal things and practices. A similar fate befalls the spiritual seeker. Instead

focal things and practices more fully, taking up Argentine tango dancing as an exemplary case in point.

Reform, then, comes through a principled restraint of technology and (re)engagement with focal practices. Borgmann argues that reforms *within* technology (and economy) complement the restraint *of* technology, in part by creating safe, hospitable environments and accessible public spaces for varied focal practices to flourish in a "republic of focal concerns."[28] If making room for and supporting what matters most is the priority, then public discourse needs to address focal concerns in a spirit of sympathy and tolerance.[29] As the ends that really matter become focal things and practices, both public deliberation over technology assessment and private-sector innovation will gain new clarity and focus. Qualitatively different criteria and judgments will restrain technology's reach and redirect its designs and uses toward service to focal concerns.[30] This is what it means to put technology in its place.

---

of engagement with and service to a religious community, one samples the spirituality de jour online knowing that many other attractive options can be summoned quickly and easily. Moreover, religious institutions unwittingly aid and abet passive religious consumption through structures, policies, and leadership styles that discourage active lay participation. What, then, is the alternative? A strong, sustained engagement with focal things and practices requires an "unlearning" of habits of interpretation and use of technological devices. Yet as Miller notes, currently our daily habits are formed under the influence of a corporate-culture industry with powerful techniques and tools at its disposal for colonizing every (pre-modern) culture. Hence the importance in *Laudato* of fashioning new-old cultural practices; as Francis observes, "a great cultural, spiritual and educational challenge stands before us, and it will demand that we set out on the long path of renewal" (LS #202). On culture as formation, see also Milbank and Pabst, *The Politics of Virtue*, chap. 8.

28. Borgmann, *Technology*, 218.

29. Borgmann is aware of tendencies toward social exclusion and cultural conflict that bedevil a communitarian politics of engagement and time focused on quality of life. A cultural and political shift in priorities from affluence to focal concerns must be complemented therefore by commitment to economic, social, and environmental justice, a commitment that includes respect for diverse conceptions of the good life. Hence a "republic of focal concerns" depends crucially on "sympathy and tolerance." In practice an Arendtian participatory politics, or what Benjamin Barber calls "strong democracy," is required to reinvigorate and in some cases reinvent institutions capable of liberally supporting a wide range of focal things and practices.

30. Feenberg argues that Borgmann tends to "essentialize" technology as an invariant pattern and hence leaves little room for the development, under changing and diverse sociohistorical conditions, of different forms of technological design and use that are more humane and environmentally sensitive, and that facilitate focal practices rather than frustrate them. On this view, Borgmann's critique of the device paradigm is necessary but insufficient; we also need to explore what Feenberg calls "reflexive metatechnical practice," which draws on science and technology studies (Latour et al.) and thus attends carefully to the changing contexts, unintended consequences, and

What Borgmann describes as a life of true wealth, or what I am calling sustainable abundance, is anchored in focal concerns yet also includes openness to the wider world and responsible, caring participation in public life. A genuinely wealthy or sustainably abundant life for all—as opposed to affluence for some—is finally what a *politics of engagement and time* must be about: "Setting aside wilderness areas, keeping rivers free-flowing, rehabilitating lakes, saving endangered species, promoting the arts, literature, and ... sciences, supporting education as a means of achieving world citizenship and the excellence of an engaging life—these measures advance the quality of life appropriate to wealth."[31] Social choices aimed at creating a good society will protect and support the focal things and practices that enable human flourishing.

Borgmann's account of the device paradigm helps to clarify and concretize Francis's view of modern technology, which is itself framed as a story of paradigms in conflict. Titled "The Human Roots of the Ecological Crisis," the third chapter of *Laudato* draws on Romano Guardini's *The End of the Modern World* and largely rehearses earlier church teaching on the errors of scientism, technocracy, and modern anthropocentrism. Francis opens with an affirmation of modern technology as an expression of human creativity and "the inner tension that impels man gradually to overcome material limitations" (CV #69, quoted in LS #102), and then restates the technological giant—moral pygmy dilemma: humanity faces grave danger "because our immense technological development has not been accompanied by a development in human responsibility, values and conscience" (LS #105). The dilemma arises historically from the fateful modern embrace of a Cartesian-Baconian paradigm of knowledge as power,[32] with the Promethean

---

surprising effects of technological designs and uses. Once propagated, technological devices can be contested, with new meanings and uses subverting capitalist efforts to de-skill workers, distract consumers, dominate nature, and colonize indigenous culture; see Feenberg, "Essentialism to Constructivism." See also Tatum, "Reform of Technology"; Tatum usefully explores the prospects for citizen involvement in creating human-centered and ecologically attuned technologies that support reengagement with focal concerns. His attention to the design process enables us to identify and criticize technology-centered designs not only through the lens of the device paradigm, but also from a range of alternative, practice-based perspectives.

31. Borgmann, *Technology*, 235. Following Galbraith, and putting the state sector aside, Borgmann views the US economy as divided into two broad market sectors, one increasingly automated and run largely by corporations, the other labor-intensive and locally owned and operated. A politics of engagement and time seeks to promote meaningful, engaging work within the latter sector. The bet here is that a push-pull combination of automation and dissatisfaction with mindless jobs will lead a significant number of workers to opt for participation in local, labor-intensive industries.

32. *Laudato* fails to mention the medieval religious roots of the Cartesian-Baconian

project of human mastery over nature allied to the false dogma of "infinite or unlimited growth" (LS #106). This "one-dimensional" paradigm now has gone global and for the most part serves elite interests. In a later chapter, Francis recalls the familiar lament over consumerism and the complicity of "mass man" with technocracy (LS #203).

Rejecting the instrumental view, Francis notes how technologies "create a framework which ends up conditioning lifestyles and shaping social possibilities along the lines dictated by the interests of certain powerful groups" (LS #107).[33] *Laudato* leans instead toward a substantive view of technology, emphasizing the difficulties of operating outside or pursuing alternatives to the technocratic paradigm's "ironclad logic" (LS #108) and relentless pursuit of power and profit (LS #108-09). Moreover, technocracy produces "specialists without spirit" (Weber) incapable of viewing the world as a complex, interconnected whole or empathizing with the weak and vulnerable. While these broad insights echoing Marx, Weber, Guardini, Mumford, Ellul, and Marcuse are valid, the distinct shape and decisive features of the "framework" and its "ironclad logic" are left unexplored.

At this point, Francis pivots momentarily to his preferred alternative, an ecological paradigm and culture shaped by it: "There needs to be a distinctive way of looking at things, a way of thinking, policies, an educational program, a lifestyle and a spirituality which together generate resistance to the assault of the technocratic paradigm" (LS #111). Liberation from the soul-deadening effects and totalizing logic of technocracy is possible, and the pope cites cooperative experiments in small-scale, sustainable agriculture as an example while also evoking moments of self-transcendence through acts of caring and the contemplation of beauty. Yet the tone here seems wistful, and soon the emphasis shifts back to the depressing undertow of the technocratic paradigm: "[T]he accumulation of constant novelties exalts a superficiality which pulls us in one direction. It becomes difficult to pause and recover depth in life. If architecture reflects the spirit of an age, our megastructures and drab apartment blocks express the spirit of globalized

---

project. As Noble argues in *Religion of Technology*, monastic pursuit of the *artes mechanicae* was understood theologically as participation in redemption, i.e., as restoration of an Edenic state of human perfection and power. Nearly every significant early modern scientist was a Christian motivated by some version of the quest to redeem humanity's sinful, fallen condition through the advance of science and technology. Modern, secular variants include the eugenics movement and transhumanism.

33. Cf. Postman, *Technopoly*: "[T]echnology must never be accepted as part of the natural order of things, . . . every technology—from an IQ test to an automobile to a television to a computer—is a product of a particular economic and political context and carries with it a program, an agenda, and a philosophy that may or may not be enhancing and that therefore requires scrutiny, criticism, and control" (185).

technology, where a constant flood of new products coexists with a tedious monotony" (LS #113). We are left with a haunting, faintly hopeful image: a mist of resistance "seeping gently beneath a closed door" (LS #113).

At the same time, Francis displays an acute awareness of the device paradigm and gestures more hopefully toward pathways beyond it. One example is the fairly detailed, balanced account of the benefits and risks associated with uses of information and communication technologies. The pope warns against habits of disengagement formed through the daily flow of online exchanges, observing that,

> when media and the digital world become omnipresent, their influence can stop people from learning how to live wisely, to think deeply and to love generously . . . Real relationships with others, with all the challenges they entail, now tend to be replaced by a type of internet communication which enables us to choose or eliminate relationships at whim, thus giving rise to a new type of contrived emotion which has more to do with devices and displays than with other people and with nature. Today's media do enable us to communicate and to share our knowledge and affections. Yet at times they also shield us from direct contact with the pain, the fears and the joys of others and the complexity of their personal experiences. (LS #47)

Francis's description of online behavior and its alienating effects reminds us of just how pervasive the pattern of technology has become without succumbing to terminal wistfulness. A simple yet powerful description of life beyond the device paradigm, and one that clearly echoes Borgmann's call for re-engagement with focal concerns, appears toward the end of *Laudato* when the pope praises those who

> show concern for a public place (a building, a fountain, an abandoned monument, a landscape, a square), and strive to protect, restore, improve or beautify it as something belonging to everyone. Around these community actions, relationships develop or are recovered and a new social fabric emerges. Thus, a community can break out of the indifference induced by consumerism. These actions cultivate a shared identity, with a story which can be remembered and handed on. (LS #232)

Making good on the promise of integral ecology requires a vivid sense of shared responsibility and sustained care for focal things, which include significant public structures and spaces.

Principled Resistance to Rapidification

More so than his predecessors, Francis recognizes the impact of social acceleration on human development and planetary life systems. Early in *Laudato*, the pope observes:

> The continued acceleration of changes affecting humanity and the planet is coupled today with a more intensified pace of life and work which might be called "rapidification". Although change is part of the working of complex systems, the speed with which human activity has developed contrasts with the naturally slow pace of biological evolution. Moreover, the goals of this rapid and constant change are not necessarily geared to the common good or to integral and sustainable human development. Change is something desirable, yet it becomes a source of anxiety when it causes harm to the world and to the quality of life of much of humanity. (LS #18)

At several points Francis states that global society must slow down or pause to consider alternatives to high-speed technological society. He warns against the idolatry of power, speed, and efficiency and urges adoption of practices and policies that respect the slower pace of natural and social ecologies. Commenting on the need for a long-range perspective and foresight in national policy-making, for example, the pope says that "time is greater than space," meaning that positioning for short-term advantage within the "space" of power politics must give way to inclusive, collaborative planning and processes aimed at promoting and protecting the "long-term common good" (#178). In other words, sustainable development requires a renewal of deliberative-democratic practices that resist the "tyranny of the moment" (Eriksen).

The fifth chapter of *Laudato* includes an extended discussion of issues raised by the proactionary-precautionary debate. Francis advocates reforms of risk-management practices based on a stakeholder model of transparent, participatory decision-making: "A consensus should always be reached between the different stakeholders, who can offer a variety of approaches, solutions and alternatives. The local population should have a special place at the table; they are concerned about their own future and that of their children, and can consider goals transcending immediate economic interest" (LS #183). Attentive to power differentials in decision-making contexts, Francis reverses the weak precautionary stance reflected in the *Compendium*. Addressing the burden-of-proof question head-on, the pope recognizes a duty to take precautionary action up front in situations where serious, irreversible harms may occur:

> This precautionary principle makes it possible to protect those who are most vulnerable and whose ability to defend their interests and to assemble incontrovertible evidence is limited. If objective information suggests that serious and irreversible damage may result, a project should be halted or modified, even in the absence of indisputable proof. Here the burden of proof is effectively reversed, since in such cases objective and conclusive demonstrations will have to be brought forward to demonstrate that the proposed activity will not cause serious harm to the environment or to those who inhabit it. (LS #186)

Development projects and new technology deployments are welcome, provided they allow all stakeholders a seat at the table, are open to ongoing reassessment and a full range of alternatives, and do not compromise human and environmental health. Importantly, the latter requirement places limits on the use of cost-benefit analysis in assessing technological and environmental risks (LS #190). In effect, Francis steers a middle course, rejecting both a reckless dismissal of the precautionary principle by technophiles and misguided attempts by some greens to use it as a blunt policy weapon for demanding a dead-stop to techno-economic progress. Along these lines, chapter 9 makes the case for a creative-middle approach in assessing technology and mediating the proactionary-precautionary debate.

Sowing Seeds for an Evolutionary Ethics of Responsible Care

Laudato concludes with a call to ecological conversion and active Christian participation in a bold cultural revolution involving a new mindset, new habits of the heart, and adoption of more intentional, ecologically sensitive lifestyles: "An integral ecology includes taking time to recover a serene harmony with creation, reflecting on our lifestyle and our ideals, and contemplating the Creator who lives among us and surrounds us, whose presence 'must not be contrived but found, uncovered'" (LS #225). Toward this end, Francis retrieves a rich kataphatic-mystical tradition of divine immanence that stresses the inter-connectedness and intrinsic value of all things and God's intimate, loving presence within all creation. While the theological vision of *Laudato* still presumes a Thomist metaphysics of created order and harmony, it also includes language congenial to process-oriented eco-theology:

> In this universe, shaped by open and intercommunicating systems, we can discern countless forms of relationship and participation. This leads us to think of the whole as open to God's transcendence, within which it develops. Faith allows us to interpret the meaning and the mysterious beauty of what is unfolding. We are free to apply our intelligence towards things evolving positively, or towards adding new ills, new causes of suffering and real setbacks. This is what makes for the excitement and drama of human history, in which freedom, growth, salvation and love can blossom, or lead towards decadence and mutual destruction. (LS #79)

Here Francis describes the universe and human history as an unfolding, open-ended drama full of possibilities for good or ill, for the realization of "freedom, growth, salvation and love" or a collective descent into "decadence and mutual destruction." From this perspective, the human vocation may be viewed as the convivial co-creation of greater value and beauty through a more profound, intimate encounter with unfolding plenitude in universal history. In this passage and others, Francis is sowing seeds for an evolutionary ethics of responsible care, the outlines of which I delineate in chapter 9.

A Promising Point of Departure

The social teachings of Benedict and Francis offer a promising point of departure for new forms of inventive, inclusive Christian praxis and moral leadership in coming decades. Part II develops a constructive, forward-looking approach to Catholic social thought and action that builds on the church's tradition and recent teaching while also suggesting possible avenues for further development. It is constructive in the sense of moving beyond both modern, totalizing discourses and postmodern, deconstructive poses toward new experiments in building sustainable community, reinventing institutions, transitioning well to a socially inclusive post-jobs society, renewing a truly humane culture, and resynchronizing natural and social ecologies for the sake of human/planetary flourishing. I also explore how Christian leaders might join with others to advance a new-distributist program through a politics of engagement and time informed by a three-pronged transformational strategy of displacement, radical reform, and disruption.

Part II

Christian Praxis and Moral Leadership in a Risky, Runaway World

6

Freedom, Hope, and the Practice of Envisioning the Adjacent Possible

> Evil does not approach us as pride any more, but on the contrary as slumber, lassitude, concealment of the "I"... It may make us so quickly contented, that any definitive fire will die down.
>
> —ERNST BLOCH

> Resisting the pervasive sense of social paralysis, the poetic imagination would nourish the conviction that things *can be changed*. The first and most effective step in this direction is to begin to *imagine* that the world as it is could be *otherwise*.
>
> —RICHARD KEARNEY

In "The City of Saba," Rumi describes a fabulously wealthy populace incapable of wonder, care, and gratitude. An evil slumber has overtaken Saba's denizens, an affliction that drains them of any definitive fire for change: "This overrichness is a subtle disease. Those / who have it are blind / to what's wrong and deaf to anyone who points it out. / The city of Saba cannot be / understood from within itself!"[1] We find Saba writ large in the results of a recent Greendex survey on environmental attitudes in eighteen countries across the planet. Conducted by the National Geographic Society and the research consulting firm GlobeScan, the online survey of 18,000

---

1. Barks, *Soul of Rumi*, 65.

people shows how the affluent within the global North are green-lifestyle laggards compared to residents in the global South. In the US, Canada, UK, Germany, and Japan, the most privileged feel the least guilty about their outsized ecological footprint and are less willing to adopt more environmentally friendly behaviors.[2] A disease at once subtle and highly communicable, this strain of affluenza is spreading to India, China, and other nations with a growing middle class. By 2030, "Asia will host 64 percent of the global middle class and account for over 40 percent of global middle-class consumption."[3]

For Rumi, there's no social remedy for the Sabanites, only the individual medicine of spiritual practices such as silence, humility, and renunciation of power and wealth. In the poem's closing lines, he implores its comfortably numb residents to "turn toward teachers and prophets who don't live in Saba. / They can help you grow sweet again and fragrant and wild and fresh and thankful for any small event."[4] How might this turning (or returning) to wisdom's teachers and traditions occur in a fast-paced technological age?

The practice of reading Rumi meditatively renews sacramental sensibility and transforms personal vision, as does the simple habit of pausing a long moment over one's meal to give thanks. Such practices expand capacities for gratitude, empathy, and imagination. As we become more capable through ongoing conversion of tasting and seeing creation's goodness and beauty, we also grow in our ability to imagine an alternative future, a world beyond Saba and the darkness it denies. We are better prepared for generative dialogue and the shared exercise of what Richard Kearney calls an "ethical-poetical imagination," or what liberation theologians call a prophetic-utopian imagination. We can envision and work toward a new and better society, a society of sustainable abundance in which people enjoy both freedom from oppression, exploitation, and marginalization and the freedom to develop their personalities fully and flourish within healthy, supportive communities.

Today's true teachers and prophets include not only high-profile figures such as Pope Francis and the Dalai Lama, but also thousands of lesser-known leaders and activists whose energies, alternative visions, and creative initiatives drive today's "blessed unrest" (Hawken) and stand in hopeful contrast to the indifference of so many. What separates the energized outside Saba from the spiritually and socially lethargic within is an

---

2. Stone, "Environmental Attitudes"; see also Crompton and Kasser, *Meeting Environmental Challenges*.

3. Rohde, "Swelling Middle." On the "rise of the rest," see Zakaria, *Post-American World*.

4. Barks, *Soul of Rumi*, 65.

ability and willingness to empathize with the victims of unjust structures and a capacity to envisage "the plausibility of the possible rather than the inevitability of the probable," as Marshall Ganz puts it.[5] Despite moments of dis-ease, many among the privileged find all-too-probable a continuation of the social-acceleration paradox: the more quickly things change now, the more they stay the same. There appears no way off the merry-go-round, and the ride is rather comfortable.

Authentic hope nurtures a productive imagining of life beyond the state of "frenetic standstill" (Rosa), while genuine freedom entails release from the false promise of a Saba-like "full life." As Borgmann argues, pursuit of a sustainably abundant life beyond the entrapments of affluence involves a principled restraint of technology and (re)engagement with focal things and practices (including spiritual practices). For socially conscious Christians, it also calls for new forms of inventive, inclusive praxis aimed at democratic renewal, social empowerment, and environmental sustainability.

This chapter aims to clarify the theological-ethical vision underlying my exploration in Part II of a constructive, forward-looking approach to Catholic social thought and action within a North American context of privilege and continuing complicity with the pervasive pattern of technology and social injustice. After presenting six brief stories of creative freedom, I explore prospects for and obstacles to systemic change as a way of introducing a philosophical perspective on the practice of envisioning the adjacent possible. I then consider theologically what it means for Christians to envision the adjacent possible of sustainable abundance for all. Here I take up the related themes of ongoing conversion, hope, and the dynamism of authentic Christian living. Finally, I offer a meditation on the human encounter with unfolding plenitude in universal history.

Exercises of Creative Freedom

Let's begin with six real-world examples of people acting collectively to create a better world. These social change stories show us what envisioning and pursuing a world of sustainable abundance looks like in practice (or better, what it looks like as praxis). The first three point toward a renewal of representative and associative forms of democracy, while the latter three offer ways in which clean technology, socially-responsible business, and effective

---

5. Ganz, "Leading Change," 211. Cf. Dussel's observation in *Ethics of Liberation*: "What enables one to situate oneself from the standpoint of the alterity of the system, in the world of everyday life of prescientific common sense, but without ethical complicity, is the ability to adopt the perspective of the victims of a given . . . system" (207).

democratic control over economic life can enhance both individual and community well-being. Taken together, they suggest the idea and ideal of sustainable abundance for all is tied closely to community-based practices designed to democratize wealth, expand economic opportunity, promote meaningful participation in political decision-making, and resynchronize natural and social ecologies.

Participatory Budgeting in Porto Alegre

In 1988, leaders and citizens in Porto Alegre, a medium-sized city in southeast Brazil, established a participatory budget process that, by all accounts, has proven successful. In briefest sum, the process unfolds as follows: each year, citizens assemble in local regions throughout Porto Alegre and deliberate in plenary sessions over priorities for the city budget. Some citizen assemblies are organized by themes with municipal scope, such as public health. These local assemblies choose delegates, who in turn gather in regional and thematic budget councils to hammer out spending priorities. Council meetings are held over a three-month period throughout the city, where citizens and leaders of civic associations discuss proposals and alternatives with delegates. Here the community grapples with the nitty-gritty of sewage lines, daycare centers, bus lines, brownfield conversions, and other issues and needs. Defensible reasons and supporting evidence matter as priorities are weighed and budget proposals are refined. During a second plenary the refined budget proposals are ratified by the local assemblies, and two delegates are selected to represent each assembly in a city-wide Participatory Budgeting Council. This body formulates an integrated, municipal budget and presents it to the mayor, who can accept it or return it to the council for revisions. Once the council and mayor come to agreement, the budget goes to the regular city council for adoption. The entire process unfolds over six months and provides tens of thousands of ordinary citizens with the opportunity to participate meaningfully in shaping the city's future.

Porto Alegre's institutional experiment in participatory democracy has shown evidence of success in several ways: (1) a massive shift in spending towards the poorest regions of the city; (2) high and sustained levels of citizen participation; (3) corruption replaced by a clean, transparent process; and (4) civic associations contributing to and growing through the process.[6] In turn, the Porto Alegre model has inspired an international movement to institute participatory budgeting processes at the city, county, and state levels. As of 2015, over 1,500 cities and institutions in North America,

---

6. For a detailed case study, see Fung and Wright, *Deepening Democracy*.

Latin America, and Europe have taken up the challenge of implementing participatory budgeting in one form or another.[7]

Passing an Anti-Corruption Law in Tallahassee

Big-money influence in US politics has grown since the Supreme Court's *Citizens United* decision in 2010; wealthy individuals and well-heeled corporations now dominate the candidate-selection process through unlimited donations to super-PACs. The 2014 mid-term elections were the most expensive in US history.[8] In response to this plutocratic trend, a bipartisan movement to end political corruption has arisen. Spearheaded by the activist group Represent.Us, the movement has succeeded in passing anti-corruption measures in nearly twenty cities and counties, including Seattle, San Francisco, and Tallahassee where two-thirds of the city's voters supported a 2014 referendum to amend the municipal charter in ways that rein in moneyed interests. In Tallahassee a left-right alliance organized by Represent.Us pushed for the charter amendments. Under the new rules, contributions to city candidates are limited to $250 per donor, each voter will receive a tax rebate of up to $25 for campaign contributions, and a new ethics board will be charged with enacting a strong ethics code that includes a conflict-of-interest policy. In the 2016 election, the first state-level Anti-Corruption Act was approved by South Dakota's voters, despite an opposition campaign funded largely by the Koch brothers and subsequent machinations by state legislators to render the new law moot. As of October 2017, Represent.Us has 700,000 online supporters and is forming local and state chapters across the country.

Growing a Social Economy in Québec

In 1996 the provincial government of Québec convened a multi-stakeholder summit to deal with long-term problems of unemployment and economic development. What made this public-policy forum innovative was the inclusion of social-movement groups, community organizations, and other grassroots associations along with the usual actors: unions and trade associations representing employers. Through deliberation the summit participants developed new policies for supporting the development of a more robust, efficient social economy in the province. Non-profits now enjoy

---

7. New models include e-participatory budgeting in Reykjavik, Iceland.
8. Mayer, *Dark Money*.

more access to credit, government grants, and indirect subsidies that enable them to provide childcare and other services (e.g., nonmedical homecare for the elderly) that meet community needs and support workers with families. The summit also led to the formation of an umbrella civil-society organization, the *Chantier de l'économique sociale*, a "network of networks" that coordinates strategies for growing the social economy. One result is innovation in institutional design: a "solidarity cooperative" (used primarily for eldercare) that meaningfully involves various stakeholders—care workers, service users, and community representatives—without creating a bureaucratic drag on service provision. The takeaway: Québec's social economy demonstrates how associational democracy can foster social learning and innovation through collaboration among all stakeholders; such collaboration enhances the quality of life for the entire community, and especially for vulnerable groups. [9]

Vertical Farming near Scranton

Once a regional powerhouse centered on anthracite coal extraction, Scranton, Pennsylvania has been in economic decline for decades. While efforts to reinvent "Electric City" as a regional technology hub have stalled, one bright spot is Green Spirit Farms' vertical farming operation, located just outside the city in Dalton, PA. In 2014 the commercial-scale farm produced its first harvest of high quality, pesticide-free, non-GMO fruits and vegetables in a former warehouse that measures 300,000 square feet. Millions of plants are stacked on vertical rows, fed nutrients by water-conserving, soil-free hydroponic systems, and lit by LEDs that mimic sunlight. The indoor facility operates year round, and its water-recycling system is highly efficient when compared to conventional, outdoor growing methods. Looking ahead, Green Spirit plans to develop other vertical farming projects using renewable energy whenever possible as well as compostable and recycled packaging. Interest in and experiments with vertical-farming designs are taking root in other parts of the US as well as other countries.[10]

---

9. Wright, *Envisioning Real Utopias*, 204–16.

10. In Jackson, Wyoming, social enterprise Vertical Harvest has built a three-story hydroponic farm to replace imported with locally grown produce.

Distributed, Renewable Energy in Hawaii

The shift from a centralized, hydrocarbon-based utility model to distributed, renewable energy systems is now taking hold in Hawaii through a market-transition process known as "cascading natural deregulation." As micro-grids powered by renewables have become more reliable and cost-effective over the past decade, demand for electricity from Hawaii's main utility has declined even as its fixed costs remain constant. With solar installations doubling each year and more users leaving the main grid, remaining customers are left to pay higher rates, which incentivizes still more people to opt for micro-grid power.

Internationally, Germany's rapid shift to locally controlled renewable energy has benefited from feed-in tariffs, a rejection of nuclear power, and other policies designed to accelerate the transition to clean energy.[11] Distributed, renewable energy projects are springing up elsewhere, and are especially suitable for countries in the global South that lack industrial-age energy infrastructures. What is happening now with rapid adoption of cheap cell phones in Africa and other developing regions could happen quickly with renewable-powered micro-grids.

Worker-Owned Cooperatives in Cleveland

By design, cooperatives democratize wealth rather than concentrate it. The Evergreen Cooperatives are a coordinated group of worker-owned cooperatives in Cleveland, Ohio committed to creating living-wage jobs for low-income residents and sustainable, democratically run workplaces. Several anchor institutions—Cleveland Foundation, Cleveland Clinic, University Hospitals, and Case Western Reserve University—partnered with city government to set up the Evergreen Cooperative Initiative in 2008. Instead of fire-sale privatization and long tax holidays to induce corporate investment—a fate that post-Katrina New Orleans and other US cities are facing—Evergreen represents a collaborative, bottom-up approach to economic revitalization in distressed urban areas. Modeled after the well-known Mondragon worker-owned cooperatives in Spain, the group includes the Evergreen Cooperative Laundry, Ohio Cooperative Solar, and Green City Growers Cooperative. They also have formed a business-development group tasked with identifying and growing new cooperative opportunities.

---

11. See Klein, *This Changes Everything*, chap. 3. At present the continuing use of lignite coal-burning plants is undermining Germany's goal of reducing greenhouse gas emissions by 40 percent by 2020 and up to 95 percent in 2050.

Worker-owned cooperatives such as Evergreen are part of a larger movement that includes co-ops in the agricultural, manufacturing, energy, insurance, banking, retail, healthcare, and other sectors. According to the University of Wisconsin Center for Cooperatives, "nearly 30,000 US cooperatives operate at 73,000 places of business throughout the US. These cooperatives own >$3T in assets, and generate >$500B in revenue and >$25B in wages."[12] In Europe, cooperative movements have thrived in the Basque country of Spain and Italy's Emilia-Romagna region. Catholics have made significant contributions to the cooperative movement, and Pope Francis is a longtime supporter of this approach to economic life.[13]

These ongoing stories of creative freedom give us a glimpse of how a vision of sustainable abundance for all can move from imagined future to reality. They point up the potential for democratic-egalitarian initiatives to take hold and grow when conditions allow and change agents seize the opportunity. Movement along multiple fronts toward systemic change is possible wherever community needs or injuries become acute, existing power structures are at least somewhat open to challenge, and a creative minority is capable of articulating and acting on a compelling vision. At various scales, an inspiring vision can activate the collective will to pursue the *adjacent possible*: desirable and viable options for pioneering new practices, advancing innovative policies, and reinventing institutions that hitherto were ignored or dismissed by many as "unrealistic" or "impractical," and thus unachievable.

Why Aren't There More Stories Like These?

The potential for democratic renewal and social empowerment is grounded biologically in evolved social instincts for freedom, fairness, and care shared by all human beings.[14] At the same time, in tension with these innate capacities is the reality of humanity's "ontological frailty."[15] We are vulnerable creatures—physically, psychologically, and socially—and hence the need

---

12. University of Wisconsin Center for Cooperatives, *Economic Impact of Cooperatives*.

13. Schneider, "Much Can Be Done!"

14. In the parlance of evolutionary theory, the desire for freedom and fairness as well as the capacity to care has survival value. Jonathan Haidt argues that freedom, fairness, and care are evolved social instincts possessed by all humans; see Haidt, *Righteous Mind*. On care and empathy as an evolved human capacity, see also de Waal, *Age of Empathy*.

15. On vulnerability as fundamental to the human condition, see Turner and Rojek, *Society and Culture*.

for stable structures, routines, and security remains in permanent tension with the impulse to risk new ventures, speak out, and empathize with others beyond our small circle. We are both inspired and unnerved by the call to exercise creative freedom in the quest for social justice. This paradox of freedom—liberating change at once desired and feared—is widely reported by community activists working with oppressed groups submerged in a culture of silence, but it applies more generally to contexts of privilege as well.[16]

For most individuals in US society the paramount reality is one of meeting the demands placed upon them by the capitalist system and its institutions, which demand above all that one be useful and productive—or else.[17] Subjectivities are shaped through cultural norms (e.g., the work ethic, the good life as maximal consumption) and the discipline imposed by socialization processes. We are prescribed to and monitored more than we care to admit, and even more so now that surveillance technologies have become ubiquitous. Moreover, as Francis notes in *Laudato si'*, the nonstop celebration of consumer sovereignty and meritocracy conceals the pattern of technology, the structures and habits of which place limits on alternative forms of lifestyle and community. Since coloring outside the lines can be costly, particularly for minority groups, many shrink from bold exercises of creative freedom. The pain, ridicule, and violence that often come with "creative maladjustment" (King) give one pause.[18]

The disciplinary codes and fear of punishment that lead most to conform are only half the story, though. In chapter 2, I described some of the social instabilities attending today's world risk society as well as the forces driving social acceleration. I noted how formerly stable, supportive institutions—e.g., family, neighborhood, state-sponsored social programs—are for many people no longer reliable. There's a growing sense that you're on your own to self-style a life and make your own way as a free agent with a "personal brand" in Market World. With technological innovation driving rapid social change and quickening the pace of life, one feels compelled to stay current with new skills, new contacts, and new opportunities for

16. The paradox of freedom is analyzed by Freire in *Pedagogy of the Oppressed*. In his foreword to Freire's classic, Richard Shaull suggests a wider application: "Our advanced technological society is rapidly making objects of most of us and subtly programming us into conformity to the logic of its system. To the degree that this happens, we are also becoming submerged in a new 'culture of silence'" (33). Shaull's summary characterization of alienation and disempowerment among the privileged echoes Borgmann's more nuanced analysis of complicity with technology's pervasive pattern, habituation to which renders many of us passive and disengaged.

17. See McKenzie, *Perform or Else*.

18. At the same time, countercultural activity more often than not gets absorbed into or co-opted by consumer society rather than suppressed.

"lifestyle enhancement." Such conditions, while exhilarating for some, generate for many a sense of insecurity and vulnerability. With no levees to hold back the flash-floods of a volatile economy—the next financial collapse, the next downsizing, the next illness that results in being let go—people become risk averse and self-limiting in their aspirations. Routines of survival and self-protection become the norm for many in a risky, runaway world. Yet all this comes at a terrible cost. Rather than reach for more, too many allow a slow diminishment of personal vision, agency, and self-efficacy—what Bloch calls the concealment of the "I"—to set in. With the soul severed from its own deepest desires and energies, it gets harder to resist social paralysis and imagine things could be otherwise.

Beliefs concerning the inevitability of the probable loom large as a limiting factor for enlisting persons into movements such as Represent.Us, Rolling Jubilee, 350.org, and other innovative projects aimed at systemic change. Social philosopher Eric Olin Wright notes how the capitalist system reproduces itself and maintains a relative stability through four means: coercion, institutional rules, ideology, and material interests. "Of the various aspects of ideology and belief formation that bear on the problem of social reproduction and potential challenges to structures of power and privilege," Wright observes, "the most important are beliefs about *what is possible*. People can have many complaints about the social world and know that it generates significant harms to themselves and others, and yet still believe that such harms are inevitable, that there are no other real possibilities that would make things significantly better, and that thus there is little point in struggling to change things, particularly since such struggles involve significant costs."[19]

Wright's reminder of how fear and self-limiting beliefs function to maintain an unjust status quo serves to sharpen the distinction between the few who grasp the power of a prophetic-utopian imagination and stand ready to exercise it with others, and the many who don't believe (for now) that "another world is possible." The many, however, are not a monolithic group, a mass of Muggles hopelessly submerged in a "one-dimensional society" (Marcuse) and thus incapable of historical agency. Under the right circumstances, ordinary people become active participants in struggles for liberation, part of a disruptive and heterogenous "multitude" (Hardt and Negri) that creatively marshals countervailing power against the forces of oppression and exploitation.[20]

19. Wright, *Envisioning Real Utopias*, 286; emphasis in original.
20. Hardt and Negri, *Multitude*. On the effectiveness of nonviolence as a method for social change, see Schell, *Unconquerable World*; also Chenoweth and Stephan, *Why Civil Resistance Works*.

As the six stories above suggest, it is in fact possible for people to come together and exercise freedom in bold, creative ways that put the lie to the oft-heard claims that there is no alternative. Even in trying circumstances, the choice to join with others in creating an alternative future rather than conforming remains. As theological ethicist John Hart reminds us, "When we use our imagination and work to concretize what we see as possible, we create our world; when we neglect our imaginative powers, or reject (out of fear of change, or fear of controversy, or fear of commitment) what they offer as possibilities, we are created by our world."[21] In sum, creative historical agency expressed through inventive, inclusive praxis can and does change the world. The section to follow explores, from a philosophical perspective, how the insight, energy, courage, and creativity required for democratic renewal and social empowerment may arise from the risky yet liberating practice of envisioning the adjacent possible.

The Practice of Envisioning the Adjacent Possible

Dismal as prospects for liberation and human flourishing may appear, the current situation in any given community or society always contains within it the adjacent possible, or what Ernst Bloch refers to as the "real-possible" (which is also the "Not-Yet-Become").[22] A broader conception of reason underlies this claim about what could come to be, one that moves beyond short-term calculations of probabilities toward a more expansive, imaginative sense of what is possible in the present moment. Commenting on Bloch, social theorist Kathi Weeks notes how "in order to grasp the present, we must not only understand its emergences from and attachments to the past, but also attempt to grasp its leading edges and open possibilities; everything real has not only a history, but also a horizon."[23] The concept of the adjacent possible implies faith in the human capacity to discern—through the marriage of reason and imagination—how the present is pregnant, as it were, with *alternative horizons*. Such horizons gesture toward a qualitatively different and better future, a future that may include broad social transformation and not simply incremental change through piecemeal reform.

For Bloch, a "Front" latent with possibility invites not flights of fancy, but rather a productive reverie oriented to world improvement. Within existing structures, change agents can envision possibilities for realizing what Bloch refers to as "concrete utopias." In contrast to abstract, fantastic utopias

---

21. Hart, *Ethics and Technology*, 149.
22. Bloch, *Principle of Hope*, vol. 1, chaps. 17–18.
23. Weeks, *Problem with Work*, 189.

that remain unconnected to historical context and remote from lived experience, concrete utopias inspire collective action along a trajectory that discerns hidden or suppressed potentialities embedded within the present situation. The exercise of a concrete-utopian imagination unlocks energy for an always partial yet significant reshaping of the social terrain—which often entails a new way of inhabiting the natural terrain as well—in ways that expand real freedom and contribute to both human and planetary flourishing. Conditions for the possibility of social empowerment are more or less open, more or less fluid, depending on a variety of factors. That leaves room—sometimes just a little, sometimes more—for pursuit of the adjacent possible.

To recall our first example, no one in Porto Alegre back in 1988 knew whether the "radical idea" of participatory budgeting put forward by Workers Party activists would work, but the proposed initiative came at a moment when, politically, enough local elites and ordinary citizens were open to the experiment. Their openness was motivated in no small measure by disgust with endemic corruption, which created political space for new ideas that previously stood little chance of gaining a hearing. (A similar dynamic has unfolded in Iceland, where participatory democracy has taken hold following the 2007–2008 financial meltdown and subsequent legitimation crisis.) For the concrete-utopian imagination, then, "the present is a fulcrum of latencies and tendencies."[24] What Kearney calls the "ethical-poetical imagination" can expand the range of what is considered possible, especially when crisis conditions appear.[25]

According to Bloch, the cognitive-imaginative act of envisioning a concrete utopia depends not only on discernment of tendencies lurking within the present but also on openness to and anticipation of the *novum*, the wholly "unexpected and transformative 'leap into the New.'"[26] What is adjacent or near often is familiar—but not always. The *novum* as unprecedented event cannot be predicted or engineered into existence. Hence the practice of envisioning the adjacent possible—a concrete utopia worthy of our hope—involves thinking "the relationship between present and future both as tendency and as rupture," as Weeks puts it.[27] Differently stated, we discern and anticipate possible new directions, yet always with the recognition that a "black swan event" (Taleb) may appear and change

24. Ibid., 196.

25. Negative possibilities also exist, as when demagogues, fundamentalist groups, plutocratic elites, and authoritarian regimes seek to exploit deteriorating social conditions through a politics of fear and resentment.

26. Weeks, *Problem with Work*, 196.

27. Ibid., 197.

everything—for good or ill. In seeking to surface and hold up what could be different going forward, a concrete-utopian imagination maintains a tension between discernible openings for change glimpsed here and now and the always surprising appearance of a *novum* that disrupts all expectations for what the future holds.

Finally, the risky practice of envisioning the adjacent possible engages our whole self and demands self-transformation over time. As Weeks notes, a crucial complement to concrete-utopian imagining is the cultivation of hopefulness, which implies an openness to personal and social changes that are transformational and hence also frightening: "Hoping as a cognitive practice may require us to think in terms of both tendency and rupture, to reconcile the future as emerging from and linked to the present, yet radically unrecognizable. But hopefulness . . . requires a great deal more: to will both (self-)affirmation and (self-)overcoming; to affirm what we have become as the ground from which we can become otherwise."[28]

What animates our hope for becoming otherwise, both as individuals and as a species? Bloch traces the utopian impulse toward human fulfillment and perfection through a wide variety of cultural forms in history. Utopian literature and drama, for example, ranges from ancient Greek comedies (Aristophanes) to medieval millenarian visions (Joachim de Fiore) to early-modern social-utopian tracts (More, Campanella) to Romantic tragedies (Goethe) to modern novels (Bellamy, Morris, Wells). These utopian aspirations from the past carry a hope content available to a hermeneutics of retrieval at the service of breaking open the present for the arrival of the New. In other words, the future buried in the past may be recovered and fused with today's dreams of a better tomorrow. Such acts of interpretation disclose a disruptive Not-Yet lurking within the present. When these buried and still undischarged futures surface, they unsettle the sediments of routine—the "great Time-and-Again," as Bloch puts it—holding otiose structures in place.

However, while the myriad wishful images present in mythology, art, religion, philosophical idealism, and visionary politics can defamiliarize things and awaken hope for something *more*, Bloch claims they cannot of themselves satisfy the desire to become otherwise in history, i.e., to transform society and, ultimately, to perfect humanity. As a Marxist, Bloch believed the science and practice of historical materialism gave humanity a genuine way forward, a concrete utopia capable of closing the gap between present suffering and future hope. The true architect Marx provides both utopian vision and a practical, scientific method for achieving it; only

28. Ibid., 203.

dialectical materialism provides sufficient grounds for a hopefulness (or enthusiasm) that drives the socialist project forward with a clear sense of direction. Past all waking dreams and wishful images of escape from death, reunion with the divine, metamorphosis, redemption from bondage, a return to one's homeland, and countless other utopian locutions, one finds a sober, scientific, and singularly powerful alternative in the Marxist utopian project, a historical project aimed squarely at building a new society freed from alienation and reconciled with nature.[29]

Such was Bloch's secular faith, one that in light of historical experience, and in principle, we cannot share. Christian realism acknowledges that every historical achievement succumbs unavoidably to ambiguity and remains provisional. Ironically, Bloch could not follow his own method by cracking open an ossified, oppressive Soviet system through the exercise of a concrete-utopian imagination that would dare to denounce the Gulag in the name of a *novum humanum* that communist ideology claimed to champion.[30] Still, Bloch's penetrating analysis of the concrete-utopian imagination helps us to grasp how the practice of envisioning the adjacent possible might awaken real hope for democratic renewal and social empowerment today.[31] Theologians have been mining insights from Bloch's philosophy of hope for some time now, as upcoming sections will show.

Discipleship and the Drama of Ongoing Conversion

For Christians, the challenge of honestly and humbly facing up to what (or who) one is and, simultaneously, audaciously desiring to become otherwise can be redescribed as the life of discipleship, understood as ongoing conversion in its several dimensions: intellectual, moral, and affective/religious. Catholic theologian Bernard Lonergan describes conversion as a lifelong process structured by an ongoing tension between two principles—one of limitation, the other of transcendence—within the developing self (or subject), who is always already a social being embedded within and responsible to others in community. These opposed but linked principles (or poles) constitute the "integral dialectic of the subject," with one pole tending toward self-maintenance or psychic stability, while the other tends toward

---

29. Bloch, *Principle of Hope*, vol. 3, chap. 55.

30. Moylan, "Bloch against Bloch."

31. Along with Levinas and Freire, Bloch plays a vital role in Dussel's model of critical praxis, specifically in the moment when a critical community of victims and those in solidarity with them jointly explore possible liberation projects and evaluate their feasibility; see *Ethics of Liberation*, chap. 5.

growth or self-transcendence. (Similarly, process philosopher Alfred North Whitehead sees the interplay between two principles—conservation and change—at work throughout the universe,[32] and we'll see in chapter 9 how Lonergan posits an integral dialectic of community shaping the unfolding of human history.)

For Lonergan, the drama of both human development generally and Christian discipleship in particular unfolds in communal contexts through moments of self-transcendence in which the *novum* of new insights, changes of heart, courageous decisions, creative initiatives, and responsible actions lead one to become otherwise, i.e., to experience conversion as growth toward authentic subjectivity.[33] At the same time, we may not rise to the challenge of becoming a more capable, creative, responsible, and loving person. This "basic sin" amounts to a denial of our deepest desires: to know and love God above all, and to know and love others (and the world we inhabit) more fully. Basic sin is rooted in a failure of the will to "be attentive, be intelligent, be reasonable, be responsible, be loving, develop and, if necessary, change."[34] In other words, basic sin "is not an event; it is not something that positively occurs; on the contrary, it consists in a failure of occurrence, in the absence in the will of a reasonable response to an obligatory motive."[35] Moreover, we are all prone to various biases that distort or limit our thinking, leave us unaware of personal weaknesses, and inhibit growth toward maturity. Crucially, there is no heroic and purely immanent self-overcoming (as Nietzsche imagined). Only a healing and transforming grace received in gratitude can release us from a condition of moral impotence into the liberty of a Christian faith commitment to "do justice, love mercy, and walk humbly with God" (Mic 6:8).

Lonergan and other Christian thinkers affirm the human capacity for concrete-utopian (or prophetic-utopian) imagining, yet with Rumi they add quickly that continual reformation of the self through spiritual practices is a precondition for being able to grasp clearly what's happening and imaginatively discern the adjacent possible. Thus Lonergan's claim: objectivity is the fruit of authentic subjectivity. We can only live authentically and bear fruit to the extent that we undergo intellectual, moral, and affective/religious conversion over time. As Holmes Rolston notes, "The task of religion is to examine the self in its relationships with the world, unmasking illusions and false cares, reforming it from self-centeredness, centering it on

---

32. Whitehead, *Science and the Modern World*, 201.
33. Lonergan, *Insight*, 469–79.
34. Tracy, *Achievement of Bernard Lonergan*, 4.
35. Lonergan, *Insight*, 667.

what is of ultimate worth . . . The religious judgment is that the self must be reformed in order to eliminate its tendency toward rationalizing, and it is just this positive contribution of worship and reflection that makes possible an unbiased rationality."[36] To which we add: moral and affective/religious conversion leads to deeper gratitude and growth in empathy. An enhanced capacity to appreciate and care enables us to understand, value, and respond to reality in a more creative, responsible, and loving manner. Rumi is right: contemplative prayer, moral introspection, expressions of gratitude, acts of service, and other spiritual practices are necessary for an exodus from Saba, for becoming otherwise.

Hope and the Dynamism of Authentic Christian Living

Theologically, openness to the *novum* is another way of speaking about Christian faith in the power of God to "make all things new." As political and liberation theologians remind us, Christian hope arises from the "dangerous memory" (Metz) of God's saving acts in history. Moments when freedom and justice flourished can be retrieved to renew hope; remembrance can become revolutionary when mingled with prophetic-utopian anticipation, especially under oppressive conditions that cry out for redress. Grounded in dangerous memory, the denunciation of present injustice prepares the way for annunciation of a kingdom to come "*on earth* as it is in heaven." With humility and courage, Christians dare to articulate and act on a prophetic-utopian vision of a qualitatively different and better society. "From first to last, and not merely in the epilogue," Jürgen Moltmann reminds us, "Christianity is eschatology, is hope, forward looking and forward moving, and therefore also revolutionizing and transforming the present."[37]

Expecting the *novum*, Christians place their faith in the promises of a loving and liberating God, creator of a "new heaven and earth." Liberating faith finds expression in the joy of communion and in a willingness to join with others in risking the adventure—despite significant costs—of social praxis aimed at realizing a more just, humane, and sustainable world. Through the eyes of liberating faith, Christians glimpse an approaching reign of God. While critical reason contributes to a social analysis that informs collective action, prophetic-utopian imaginative thought generates an inspiring vision that ignites hope and fires commitment, and at the same time saves one from falsely identifying particular historical projects with God's reign. As liberation theologian Gustavo Gutiérrez points out,

36. Rolston, *Science and Religion*, 31.
37. Moltmann, *Theology of Hope*, 15.

> Christian hope ... keeps us from any confusion of the Kingdom with any one historical stage, from any idolatry toward unavoidably ambiguous human achievement, from any absolutizing of revolution. In this way, hope makes us radically free to commit ourselves to social praxis, motivated by a liberating utopia and with the means which the scientific analysis of reality provides for us. And our hope not only frees us for this commitment, it simultaneously demands and judges it.[38]

The continuing vitality of Catholic social thought and action requires ongoing conversion and commitment to a liberating faith that is not afraid to exercise a prophetic-utopian imagination and fashion new forms of inventive, inclusive praxis dedicated to democratic renewal, social empowerment, and environmental sustainability. As we saw in chapter 3, Paul VI acknowledged the value of a concrete-utopian imagination for Christian praxis. Reflecting on the potency of utopian protests, the pope noted how

> this kind of criticism of existing society often provokes the forward-looking imagination both to perceive in the present the disregarded possibility hidden within it, and to direct itself towards a fresh future; it thus sustains social dynamism by the confidence that it gives to the inventive powers of the human mind and heart; and, if it refuses no overture, it can also meet the Christian appeal. The Spirit of the Lord, who animates man renewed in Christ, continually breaks down the horizons within which his understanding likes to find security and the limits to which his activity would willingly restrict itself; here dwells within him a power which urges him to go beyond every system and every ideology. (OA #37)

Past all secular utopias, Marxist or otherwise, Christians dare to imagine and hope for the *novum*, for the surprising and joyful arrival of God in their midst. The dynamism of authentic Christian living has its source in the animating power of God, the Spirit of the Lord who urges disciples to "go beyond every system and ideology." Ultimately, what inspires all prophetic-utopian visions, and every action directed toward making them a reality, is a divine mystery encountered as a tremendous, fascinating *novum* within an unfolding plenitude that defies description. In what follows I invite readers to move into the imaginary for the sake of envisioning sustainable abundance at the widest angle possible.

---

38. Gutiérrez, *Theology of Liberation*, 238.

Our Encounter with Unfolding Plenitude in Universal History

One promising way to make theological sense of the collective quest for sustainable abundance is to place it within the widest of contexts, namely the human encounter with unfolding plenitude in universal history. To gain our bearings and find our way in such a vast expanse, we need some sense of where we are as well as where we might be going. Let us join others strolling down a historic path near midnight and see where it leads.

In Berlin these days, pedestrians from far and wide walk a lighted pathway at all hours where the Wall once stood and curfews were imposed. Rigidity has given way to fluidity. This democratic promenade invites diversity. In a space of freedom, a clearing where many stroll and some dance, few mourn the death of ideologies and grand visions. A final stage or end to history? Today, such thoughts appear absurd to most, if not dangerous.

Still, as midnight approaches, in the evening breeze a whisper may be heard: What next? Where to now? The questions may be ignored, but they won't go away. One way or another, our collective search for a sense of direction in the movement of life on the home planet continues. Compelling narratives will capture the imagination of millions and set much in motion as personal stories fold into larger ones. And with motion comes friction. Short of the Omega Point, story wars and the political-military struggles they bring to life appear inescapable. For clannish, storytelling creatures we are, our life together a messily staged affair. And so, come daybreak the call will come to take the world stage and perform the scripts we ourselves pen.

Near midnight, as we stroll along freedom's path, our gaze may wander skyward. Then suddenly and without effort we find ourselves floating, feather-like, far above all the familiar markers anchoring perception. We glance back to admire the glistening blue jewel from whence we came. As we soak in earth's immense beauty—and fragility—from afar, we find we've never felt so close to it until now. At this moment the voice of the earth bids us return, lest we perish in the darkness or succumb to vertigo. Yet even as the home planet beckons we hear another voice, more distant yet no less compelling. We sense the possibility of contact with other actors dwelling somewhere amidst the billions of lights all around us. Now here is a stage and a stage call like no other! How can we resist? Our eyes turn toward what lies far beyond us, and with this gaze we gradually melt into a vast, sidereal ocean. A flood of wonder leaves us speechless.

Lingering a moment, we find ourselves weighing these two calls. We feel the gravity and responsibility of the one calling forth our care, and the expansiveness of the other bidding us to dare. Our heart wants to say yes twice—and doubly rejoice. For the only true option is to embrace both yearnings fully, to care for roots, soil, air, and water while also striving to take wing into deep space someday.

These two desires reflect who each one of us is: an evolving body-brain-heart nexus—an embodied soul—impelled toward greater knowing and loving in freedom and in communion with others. Possibilities unveiled by imagination as well as realities newly disclosed to intellect bid us to reach further into the within and without of things with growing insight, sensitivity, and awareness. What is more, the energy both to care and to dare seems activated by an unseen attractor. Our striving is continuously prompted by God, our absolute future. Our reaching inward and outward is in response to an alluring call, a call to become otherwise. Despite our frailties and fears, all the while and at every turn we are drawn forward by the promise of *more*—much, much more—to come.[39]

Every act of faith, hope, and love—every responsible, creative action—participates in and contributes to a more profound and intimate encounter with a plenitude that precedes and exceeds us infinitely, drawing us forward within the vast arc of its dynamic unfolding. Faith and hope open us to the transcendent dimension of existence in the mode of promise, of movement toward the *novum* of greater value and beauty, while love makes possible the experience of a deeper communion, a closer mutual presence, a more profound unity-in-difference. God is always already lovingly present within every pore, every corpuscle, every wave-particle of an immensely vast and evolving cosmos, present as the One who continually calls the Many to reach for and bring to realization through artful, co-creative action ever more intense, vibrant, and life-enhancing forms of value and beauty.[40] (And the many beyond this solar system may outshine us already!)

Our collective quest for a sustainably abundant life on the home planet occurs within this dynamic and unfolding superabundance. We inhabit a cosmic theater rich with possibilities, a truly grand playhouse in which every human project—however shaped and at whatever scale—remains provisional and hence also open in principle to *something more*, open to the experience of *epektasic excess* (as Gregory of Nyssa grasped). Through a more profound and intimate encounter with unfolding plenitude—the creativity of God—a *novum humanum* may come into being. The ancient vigil—with candles lit upon a hill, eyes lifted heavenward, and chants of

---

39. Cf. Haught, *Making Sense of Evolution*.

40. Cf. the creation theology and "counter-ontology" of Christianity articulated by Milbank in *Theology and Social Theory*, 429–30.

maranatha—powerfully expresses this hope, as has every Easter vigil henceforward.

Today the caretaking call beckons us as never before; we sense the gravity of protecting the earth more acutely than ever in light of climate change, the sixth great extinction, and other alarming trends. At the same time, the adventurous among us can see the day coming when the global theater we are now remodeling will prove too small a stage. We are a curious, dynamic lot in the vast, unfolding plenitude that is the play of God. Hence the current venue where the human drama unfolds will in time become a staging ground—albeit a cherished one—for the expansion of the human presence into the Milky Way and perhaps beyond.

Amidst unfolding plenitude, our vocation and comprehensive purpose is the convivial co-creation of greater value and beauty in an evolving, expanding cosmos. Human authenticity is realized in and through creative, responsible participation in divine creativity. This historical moment calls for unprecedented creativity and energy directed to earth-caring even as the quest to explore a far wider world continues to inch us onward and outward. Going forward, we do well to embrace and honor both our desire to love fiercely what's given here and now on the home planet and our desire to discover what lies out ahead in the galactic commons.[41] Most of all, we must desire communion and not simply contact with the myriad others we encounter, both here and beyond.

Touching Down and Looking Ahead

Time and gravity no longer suspended, we drift down from our reverie to the concrete walkway—and to the daunting challenges ahead. Perhaps we now stand a bit further outside Saba, more awake and ready to resist its siren calls, and more attentive to those it leaves by the wayside. On freedom's path once more, we wonder what unexpected turns history may take, what black swans may appear. Can we let go our fears and comforts and together take up the risky, liberating practice of envisioning and pursuing the adjacent possible? In this moment near midnight, dare we ask the question: Where do we go from here?

---

41. Cf. Hart, *Cosmic Commons*; see also O'Meara, *Vast Universe*.

7

Pursuing the Adjacent Possible of Sustainable Abundance for All

> We are at that very point in time when a 400-year-old age is dying and another is struggling to be born—a shifting of culture, science, society, and institutions enormously greater than the world has ever experienced. Ahead, the possibility of the regeneration of individuality, liberty, community, and ethics such as the world has never known, and a harmony with nature, with one another, and with the divine intelligence such as the world has never dreamed.
>
> —DEE HOCK

> Glimpsing the vision of a flourishing human future is life changing.
>
> —MARTIN SELIGMAN

Where freedom's path may lead in coming decades is, of course, an open question to be answered by the personal and social choices we make in light of emerging trends and conditions. An evolving, contingent, and complex world situation can take a dramatic turn, as when black swan events occur. No theory of systemic change—historical-materialist, progressivist, or otherwise—can predict with confidence the shape of things to come. Moreover, the process of hyperacceleration within which all of us are caught up renders the deliberative, time-intensive task of visioning and strategizing more difficult.

That said, we need not succumb to the incessant demands of a fast-paced present nor fly blind into the future. The limits of human foresight, irreducible indeterminacy, and obstacles to personal and social change notwithstanding, a prophetic-utopian imagination rightfully poses the eco-social question regarding the good life and the how of human/planetary flourishing. Strengthened by a liberating faith in God's transforming grace, we dare to envision and pursue the adjacent possible of sustainable abundance for all. Our vocation and comprehensive purpose on the home planet, we have said, is the convivial co-creation of greater value and beauty. At this historical juncture, I suggest, our response to the eco-social question calls for a politics of engagement and time, which prioritizes (re)engagement with focal concerns and cultivates a long-range, strategic vision that frees us from the tyranny of emergency and inspires inventive, inclusive forms of Christian praxis and moral leadership along many fronts.

In this chapter I explore further the conditions and possibilities for systemic change in the early twenty-first century. After taking stock of several macro-trends shaping the current situation, I offer a promising if partial sketch of what a diverse, hybrid economy and socially inclusive post-jobs society geared to sustainable abundance for all might look like in coming decades. I then take up Bloch's invitation to retrieve the future buried in the past by considering how the Catholic distributist tradition provides insight and inspiration for a new-distributist program befitting our time. Next I present a notable example of institutional change as a preface to an outline of a broad, three-pronged strategy for democratic renewal, social empowerment, and environmental sustainability in the US context. I close with a few thoughts on the type of leadership required to move things forward.

Rising Waves, Rising Debt

Cornel West is fond of saying that social ethics begins with Marvin Gaye's question: What's going on? Before turning to possibilities for systemic change, we need to consider briefly how five big waves—globalization, increasing environmental stress, exponential technology development, population growth, and urbanization—are shaping the planet's future, taking note of the ways in which these complex phenomena generate systemic risks.

Both globalization and the ecological crisis can be interpreted as unfolding stories about the risks and consequences of rising debt. In this sobering account, we find that accumulating debt—global-financial and environmental—threatens to wash away hopes of sustainable abundance for this generation and those to come. For its part, exponential technology development may prove the wildest wave of all, holding both great promise and grave peril, while the steadily rising waves of population growth and urbanization highlight both the human consequences of global inequality and ecological limits to growth. Along with the specter of technological unemployment, rising global-financial and environmental debt not only heightens risks for growing numbers of people, but also contributes to a gradually deepening legitimation crisis for ruling institutions, a long-term crisis that opens up opportunities for social movements and civil-society actors to articulate and act on the adjacent possible of sustainable abundance for all. Let's examine these rising waves a bit more closely, beginning with a few observations on globalization and debt.

As the balance of power continues to shift toward the East, multinational corporations remain the most powerful economic actors on the world stage, while neoliberal policies largely govern the global economy. At present, China is embarked on an aggressive program to extend its economic power and geopolitical influence throughout the developing world, e.g., through massive foreign investment in resource-rich countries, the recent establishment of the Asian Infrastructure Investment Bank, and expanded role at the United Nations. Meanwhile, during the past decade the debt crisis, usually associated with highly indebted poor countries, has become a global phenomenon constraining both public and private spending and investment. A growing aggregate or total debt load—call it simply the global debt—is a risk factor threatening the stability of the world financial system.

As of 2014, government debt represented 29 percent of the world's total debt, which also includes household (20 percent), corporate (28 percent), and financial (23 percent) debt. From 2007 to 2014, global debt climbed to $199 trillion, increasing by $57 trillion while outpacing global GDP growth. During this period, government debt exhibited a 9.3 percent compound annual growth rate—significantly higher than other debt categories—climbing from $33 trillion to $58 trillion, with 75 percent taken on by advanced industrial countries. Japan and a number of EU nations—not just Greece—are burdened with heavy debt loads, and high public debt-to-GDP ratios in these countries are projected to increase even further in coming years.[1] By June 2017, total global debt had reached a record-high 327 percent of world GDP.

Under these conditions, governments find it more difficult to fund education, health care, pensions, and social services at desirable levels. Where pressure mounts to cut government services and reduce public-sector

---

1. McKinsey, "Debt and Deleveraging."

employment, counterpressure builds up and boils over into the streets. We can expect the recent turmoil in Greece to recur in other countries as fiscal crises erupt and aroused publics attempt to salvage what remains of the social-democratic project in the EU.[2] Moreover, as populations in developed countries age and worker-to-retiree ratios shrink from 3:1 to 1.5:1 (1:1 in Japan) by the 2030s, growing pension liabilities will place even more pressure on state budgets. Meanwhile, private debt-to-GDP ratios in the US, UK, Japan, France, Germany, and China also have grown dramatically since the early 1990s, a trend that constrains consumer spending and limits the productive investments required by individuals and businesses to sustain themselves over time. Despite the global economic recovery, many lower- and middle-income households as well as students continue to struggle with high debt burdens.[3]

Budget-constrained governments also are limited in their efforts to cope with mounting "environmental debt," a term coined by economist Nicholas Stern, author of a British government-commissioned review of the economic impact of climate change. Released in 2006, the Stern Review Report characterized climate change in economic terms as a major market failure and asserted that failure to pay the up-front costs of mitigating climate change—a cost estimated at one percent of global GDP—would in the near future cost global society anywhere from 5 to 20 percent of global GDP.[4] In addition to the high cost of inaction on climate change, global society is consuming its natural capital. Business as usual translates into fresh-water shortages, collapse of world fisheries, continuing massive release of toxins, and a "sixth great extinction" leading to the loss of perhaps half of all species by century's close. This squandering of natural capital puts future generations in peril of inheriting a harsh, diminished, and inhospitable planet.[5] Pentagon strategists now consider the social dislocation, political instability, and growing likelihood of "resource wars" (Klare) triggered by climate change a major national security issue.

Mounting global-financial and environmental debts are related risk factors, and each competes with the other for elite attention. Both are systemic risks—i.e., risks to the entire system's functioning—which exhibit three characteristics: vulnerability to modest tipping points that combine

2. Cutting across and complicating these economic issues is the political and cultural question of immigration in the EU, a question now coming to a head with the Syrian refugee crisis.

3. Vague, "Private Debt Crisis."

4. For a summary and analysis of the Stern Report, see Larkin, *Environmental Debt*, chap. 2.

5. See the engaging work of Kolbert, *Sixth Extinction,* and McKibben, *Eaarth*.

indirectly to produce large failures; risk-sharing or contagion, with one loss triggering multiple losses; and hysteresis, where the system is unable to return to equilibrium after a shock and a "new normal" results.[6] In coming years, global-financial and environmental debts may interact in ways that trigger a downward spiral into economic and environmental collapse. According to World Economic Forum (WEF) researchers,

> Continued stress on the global economic system is positioned to absorb the attention of leaders for the foreseeable future. Meanwhile, the Earth's environmental system is simultaneously coming under increasing stress. Future simultaneous shocks to both systems could trigger the "perfect global storm" with potentially insurmountable consequences.[7]

Moreover, accumulating global-financial and environmental debts are not the only systemic risks on the horizon. The WEF's Global Risks 2013 study ranks "fiscal crises in key countries" as risk #1, "structural unemployment" at #2, and two environmental risks—"water crises" and "failure of climate change mitigation and adaptation"—at #3 and #5, respectively. Structural unemployment is linked to accelerating automation, which threatens to deskill and render redundant many millions of workers across the globe in coming decades (see chapter 8). Rising structural unemployment will aggravate the problems of inequality (risk #4 on the WEF list above) and debt, thus heightening the risks of political instability and social breakdown. In the global South, where population growth rates are highest, millions will continue to flood urban slums as arable lands become degraded or unusable from more extreme weather events (e.g., longer, more severe droughts), automation of agri-business continues apace, and land ownership becomes more concentrated and foreign-controlled. The huge shortfall in decent jobs already plaguing low-income countries will grow even worse unless alternative pathways can be forged.

A Partial, Promising Sketch

The findings of systemic risk analysis—whether conducted by the Intergovernmental Panel on Climate Change, WEF, or the Pentagon—render more urgent the search for new approaches to sustainable development, tipping

---

6. Goldin and Mariathasan, *Butterfly Defect*, chap. 1.

7. World Economic Forum, *Global Risks 2013*. Long a bastion of neoliberalism, the WEF has in recent years become a key site for debates between neoliberals and ecological modernizers. The report referenced here reflects a strong tilt toward the latter position.

the scales toward an ecological-modernization agenda while also creating more space for green-communitarian voices to be heard. Under the banner "Another World Is Possible," activists of many stripes gather in various venues to explore alternatives to neoliberal globalization.[8] At the local, metropolitan, and provincial/state levels, collective exercises in building shared vision occur through deliberative-democratic processes, e.g., participatory budgeting, inclusive planning for social-economic development, civic environmental councils, town meetings, climate-change compacts, and citizens' conferences. These exercises in associative democracy encourage social learning as lessons learned in one situation inform efforts elsewhere. They also contribute to and are informed by a broader, ongoing conversation focused on long-term, strategic considerations and attempts to articulate and act collectively on a broader vision of social transformation. Such a vision, open to revision as conditions change and new data appears, can inspire and inform dialogue, collaboration, and collective action on a wider scale and along a number of fronts.

Here I join that wider conversation and offer to sketch out several elements of what a diverse, hybrid economy geared to sustainable abundance for all might look like in coming decades. By "diverse, hybrid economy" I mean a more decentralized, pluralist economic system operating within planetary boundaries and characterized by a mix of capitalist-market, state, and civil-market sectors (with the latter preferred and protected in policymaking).[9] Rather than concoct a grand blueprint (always a bad-utopian move), a partial sketch has the value of stretching our imaginations and inviting us to ask "What if . . . ?" and "How might we . . . ?" questions that tend to move dialogue and collaboration forward in promising directions. For socially conscious Christians in particular, I present this sketch as a stimulant for prophetic-utopian thought.

Here, then, are four elements to ponder:

A Twenty-First-Century Smart Infrastructure

While exponential technologies accelerate automation, they also promise to yield dramatic productivity gains while helping to reduce humanity's

---

8. I have in mind here the many different generative dialogues that occur in alternative venues, from World Social Forum and New Economy Coalition events to Occupy encampments to openDemocracy discussion threads online.

9. On the notion of a diverse, hybrid economy, see Wright, *Envisioning Real Utopias*, chap. 5; also Gibson-Graham, *Postcapitalist Politics*. Under "civil market economy" I would group a diverse range of not-for-profit economic activities, which include the cooperative, social (or sharing), and care economies.

ecological footprint in locales where complementary shifts in manufacturing practices, urban-suburban design, consumption patterns, community life, and social norms take hold. According to Jeremy Rifkin, the internet of things (IoT) has the potential to

> connect every thing with everyone in an integrated global network. People, machines, natural resources, production lines, logistics networks, consumption habits, recycling flows, and virtually every other aspect of economic and social life will be linked via sensors and software to the IoT platform, continually feeding Big Data to every node—businesses, homes, vehicles—moment to moment, in real time. Big Data, in turn, will be processed with advanced analytics, transformed into predictive algorithms, and programmed into automated systems to improve thermodynamic efficiencies, dramatically increase productivity, and reduce the marginal cost of producing a full range of goods and services to near zero marginal cost across the entire economy.[10]

Along with the build-out of renewable-powered micro-grids and smart transport systems, the IoT will function as a powerful general-purpose technology for a twenty-first-century infrastructure characterized by distributed energy (as in Hawaii and Germany), highly efficient logistics featuring electric vehicles and modernized rail- and subways, and a global communications network marked by ubiquitous connectivity.

The envisioned build-out of a fully-digitalized and highly-automated infrastructure raises at least three questions: What material resources are required, and who controls sourcing and distribution? Who pays for and owns the new infrastructure? What purposes does it serve, and for whom?[11] Regarding material flows, the use of oil as feedstock for creating advanced materials is a promising new development in efforts to reduce the carbon footprint of manufactured goods (e.g., building materials, vehicles) and construction activities.[12] Financing, ownership, and control over new infra-

 10. Rifkin, *Zero Marginal Cost Society*, 11.

 11. The last two questions are addressed by Rifkin in *Zero Marginal Cost Society* and *Third Industrial Revolution*.

 12. Doherty et al., "Oil and Gas." Even if advanced materials derived from oil are able to scale, the materials and sourcing question remains problematic, given the importance for the foreseeable future of rare earth elements, tantalum, platinum, copper, gold, and other minerals used for industrial and emerging technologies. One issue is conflict minerals in the Congo and elsewhere, another the economic and geopolitical implications of China's current monopoly on rare earths (95 percent of known reserves are in China). A third issue is the considerable environmental impact of mining operations, most of which are located in the global South. On these issues, see Jacobson and

structure needs to be public, with state and national governments playing a subsidiary role in support of local-regional and metropolitan initiatives coming out of participatory-democratic planning processes.

A twenty-first-century smart infrastructure will bring with it new risks and vulnerabilities, including threats to autonomy, privacy, security, and political freedom.[13] Battles to control and shape the design and implementation of the IoT are a major front in the struggle for democratic renewal and social empowerment, and so let's next consider a proposal for democratizing a growing cyber-economy.

New Digital Network Designs for Democratizing Wealth

As digitalization becomes ubiquitous the source of wealth is shifting to idea generation and control of information. Organizations such as Google, Facebook, Amazon, Walmart, and the NSA control and exploit what Jaron Lanier calls "Siren Servers": highly powerful computer resources that collect and analyze Big Data and thereby provide their owners with immense competitive advantage and surveillance capabilities. To create a world of sustainable abundance for all, we must address the problem of extreme information asymmetry and the velvet-glove domination of Siren Servers by developing new digital platforms and network designs.[14]

One possibility is Lanier's own ambitious proposal for a humanistic digital economy. Lanier grounds his alternative model in a simple yet profound observation: information is just people in disguise. Unfortunately, current network designs fail to recognize and acknowledge this fact, and consequently all the valuable data shared by many millions of users is "off the books." According to Lanier, "popular digital designs do not treat people as being 'special enough.' People are treated as small elements in a bigger information machine, when in fact people are the only sources or destinations of information, or indeed of any meaning to the machine at all."[15] While Siren Server organizations reserve exclusive rights to exploit information flows, over time the scope of economic opportunity and privacy narrows for the many.

Delucchi, "Plan to Power" and "Providing All Global Energy."

13. See Shneier, *Data and Goliath*; Howard, *Pax Technica*; and Angwin, *Dragnet Nation*.

14. Cf. the critique of digital platform monopolies in Rushkoff, *Throwing Rocks at Google*, 82–93.

15. Lanier, *Who Owns the Future?*, 8.

Lanier's proposed solution would require a new social contract as well as changes in technology and law. The legitimacy of global cyber-economic activity will depend on a widespread belief in the basic fairness of digitalized, automated systems.[16] In the future, people must see how they can participate meaningfully—and with a reasonable chance of success—in the cyber-economy. To institutionalize a new social contract, Lanier suggests a revamping of network designs to allow for two-way communication online and reliable tracking of the provenance of data. In other words, everyone will know the source of data, and a sophisticated accounting system will track every use of it by others and process payments to the data's owner. The days of anonymous copying and ubiquitous mining of data will end, and a new regime of accountability and transparency will enable billions of IoT users to participate in the digital economy as paid contributors rather than unrecompensed "sharers" of commercially valuable information. Accordingly, society will need to establish a new commercial right to own and control one's own information, and here information includes anything from personal data gathered from wearable tech, online purchases, and other routine cyber-activity to various forms of intellectual property (e.g., a design file) uploaded to the cloud.

Under Lanier's scheme, individuals would receive compensation (micro-payments in most instances) from any person or organization accessing their data/IP either for consumption or the creation of added value. For example, if a fashion designer uploaded a new design file for a stylish pair of pants that, say, tomorrow's advanced 3D printers could fabricate quickly anywhere, then an automated system would charge the accounts of all those who access and use the designer's IP (unless, of course, she decides to share the file for free under a Creative Commons license).

While Lanier's proposal faces considerable practical hurdles and raises some disquieting questions,[17] his critique of Siren Server domination is

16. The question of equal access to the information commons through net neutrality and a closing of the digital divide—while not addressed by Lanier—is a crucial component of any new social contract.

17. A number of theoretical, practical, and moral-cultural difficulties attend Lanier's proposal, not least of which are the technical, economic, and political challenges of a large-scale switch-out of existing network designs. Lanier suggests that enlightened self-interest might lead Google and other Siren Server organizations to self-organize around a transition to a new network design. I leave it to IT experts to debate whether the current network designs embedded in the Internet and World Wide Web have attained irreversible "technological momentum"; for historical examples and analysis of this phenomenon, see Hughes, "Technological Momentum." On the ethical and cultural front, Lanier's proposal seems to be in tension with an Internet ethos of free sharing and open-source software development. With digitally mediated creation and sharing of information, knowledge, and culture (all non-rival goods, in the parlance

widely shared. Researchers at the Institute for the Future, for example, have developed a set of design principles for building "positive platforms" that "embed protocols and practices designed not only to deliver profits to their owners and investors but also to help maximize incomes and create positive work conditions for those whose livelihoods depend on them."[18] Alternative and quite sophisticated designs for distributed social networking (e.g., Synereo) and digital platform cooperatives[19] are coming online. We can expect these and other promising, independent cyber-initiatives to multiply—and some to go viral—in coming years as a key component of a democratic culture and global network society.

Distributed Manufacturing and Self-Provisioning

A twenty-first-century smart infrastructure will create the material conditions for a flowering of smaller-scale and more labor-intensive enterprises (e.g., Evergreen Cooperatives), many of which will benefit from new, affordable DIY technologies such as 3D printers. With economic security provided by a universal basic income, local entrepreneurial activity in manufacturing (and other sectors) will increase significantly. The IoT, distributed energy, ubiquitous connectivity, and basic income will contribute to a flourishing of local manufacturing both in the global South and in the advanced industrial world. Financing and production of housing, appliances, electronics, furniture, apparel, toys, and many other products will not require global supply chains and the likes of Bank of America, Toll Brothers, GE, Foxconn, Ikea, Hanesbrands, Ford, and Mattel. Rather, with the continuing "return of the micro," small and medium-sized enterprises in local-regional economies will generate quality goods and services at affordable prices, while corporate behemoths such as Walmart either learn to partner with local producers or risk going bankrupt. Economies of scale will come to matter less than dense, cooperative networks making up vibrant local economies. Capital

---

of economics), do we want a cyber-system that tracks and price-tags everything? In a world of sustainable abundance for all, won't we have the technical means for efficient and widely distributed social production of goods and services as well as ubiquitous sharing of ideas, art, games, and other products of the mind and imagination? Perhaps copyleft and Creative Commons licensing will protect and nurture an online sharing ethos existing side by side with digital markets designed to democratize wealth. As Boyle argues, we now need a cultural environmentalist movement to protect the commons of the mind from the new IP enclosures; see Boyle, *Public Domain*.

18. Fidler and Gorbis, "Prosperity by Design," 31.

19. On the promise of digital platform cooperatives, see Rushkoff, *Throwing Rocks at Google*, 215–23.

requirements will drop, while public and community-based financing becomes more widely available as pension funds, universities, charities, credit unions, community banks, and foundations divest from fossil fuels and reinvest in a new economy geared to sustainable abundance for all. Moreover, much of this distributed, DIY manufacturing will take the form of community-supported self-provisioning as people spend less time in formal employment and more time helping each other learn how to produce what they need and want for themselves in a post-jobs society.[20]

Both neoliberal privatization and big-development projects sponsored by national governments have fallen short, especially in the global South. For the tens of millions leaving the countryside and crowding into Latin American, African, Middle Eastern, and Asian cities, formal employment options will remain limited or out of reach, even among those who receive an education. With dreams deferred close by, many young men and women will light out for Europe or North America, while some will succumb to the propaganda of a Boko Haram or ISIS. Yet precarious informal-sector work, emigration, and extremism need not be the only options. Instead of chasing after foreign investment or attempting to stimulate job growth through massive public spending, national governments can partner with state/provincial and municipal authorities to encourage the rapid distribution (e.g., through interest-free loans and subsidies) of new DIY technologies and creation of community-based education and production centers that foster "new work" oriented to self-provisioning and small business creation.[21] They also can expand on the demonstrated success of basic income pilots and programs in Brazil, Namibia, India, Kenya, and other developing countries; these cash-transfer schemes provide the necessary means for individual and community self-development.[22]

Local, Sustainable Food Production

Sustainable abundance in food production will come through reengagement with farming as a focal practice rather than a big business.[23] Agro-ecology, which combines local knowledge with scientific method and appropriate technologies, will supplant unsustainably large, hydrocarbon-based monoculture methods. Innovations such as vertical farming will complement

20. Schor, *Plenitude*, chap. 4; also Schor and Thompson, eds., *Sustainable Lifestyles*.

21. Bergmann and Staehelin, *Starting with New Work*.

22. For a balanced discussion of basic income experiments in the global South, see Van Parijs and Vanderborgt, *Basic Income*, 138–40; also Hanlon et al., *Just Give Money*.

23. Thompson, "Farming as Focal Practice." See also the writings of Wendell Berry.

the resurgence of smaller-scale, community-based agricultural models, enabling urban and suburban areas to achieve greater food security through diverse, local sourcing. As the example of Green Spirit Farms shows, vertical farming's indoor operations can produce high-quality, pesticide-free, non-GMO fruits and vegetables year round. Food supplies will be more secure, because production can continue even in the midst of extreme weather conditions. Highly water efficient, vertical farming will become more attractive as fresh water sources dry up and weather patterns change due to climate change.[24]

These four elements suggest pathways for creating over time a diverse, hybrid economy in which wealth, economic opportunity, and decision-making power are more widely distributed.[25] They complement each other, forming a promising if partial picture of the adjacent possible across society. Importantly, a shift to sustainable food production and transition to a smart, post-carbon infrastructure will enable nations and regions to achieve higher levels of energy independence and food security, making large military expenditures and foreign interventions less tenable. In the US context, the envisioned transition to a more diverse, hybrid economy is consistent with what retired military leaders Mark Mykleby and Wayne Porter call a "new grand strategy" for national security based on sustainability challenges and community-development initiatives within the country rather than imperial notions of "full spectrum dominance" overseas.[26]

24. Despommier, *Vertical Farm*. A tension clearly exists between traditional farming's older methods and more intimate relationship with the land and its local flora and fauna, on the one hand, and the more artificial, technologically dependent techniques deployed in vertical farming's indoor operations. Recognizing with Borgmann the practical need for access to a spectrum of technological designs and uses—ranging from the latest high-tech devices to premodern tools—seems helpful here. While vertical farming removes its practitioners from direct, sensuous contact with the land, it does support engagement with the local community. The point can be generalized: an important practical question for a politics of engagement and time is the envisioned mix or ensemble of land uses and technologies within a neighborhood, town, city, and metro region. In some ways the twenty-first-century smart infrastructure described above seems likely to perpetuate the device paradigm. For example, tomorrow's electric, self-driving cars promise to be safer, more energy efficient, and less polluting, yet they will still amount to technological devices in Borgmann's sense. More hopefully, urban-suburban planning that combats sprawl may open up viable and more engaging forms of mobility (e.g., biking, walking, mopeding). Another example is the mix of tools and designs taken up by the new self-provisioning pioneers. How much does one rely on increasingly sophisticated 3D printers, the use of which likely will result in the erosion of manual skills associated with older tools?

25. Community-based alternatives also exist for education, health care, and other sectors.

26. Mykleby, *New Grand Strategy*.

A New Distributist Program

In chapter 3, I noted the contribution of British distributists to social Catholicism. Eric Gill, an important figure in this small yet vital movement, once quipped: "To abolish the proletariat and make all men owners—and to abolish mass-production and return to a state of affairs wherein 'the artist is not a special kind of man but every man is a special kind of artist'—that would be a revolution."[27] The distributist revolution Gill and other Catholic writers and activists envisioned a century ago remains relevant today. It's one of the fine threads from tradition we need to take up, though a new distributist weave will differ in significant ways from the earlier fabric.

Distributists then and now argue that abandonment of Catholic social wisdom has had tragic consequences for modernity: loss of freedom and widespread insecurity, excessive consumption within a sensate culture blind to its own atrophied spiritual sensibilities, normalization of a perpetual war economy, degrading welfare dependency, and systematic environmental degradation. The distributist tradition offers a powerful critique of modern ideology and industrial society. We can affirm its commitment to subsidiarity and corresponding rejection of concentrated power, whether corporate or governmental. We can appreciate its communitarian ethos as well as its insight into the close connections between substantive freedom, effective control over the means of production, and widespread ownership of productive property. Distributists are correct in claiming that modern society's systemic ills arise from a tragic distortion of the integral scale of values upon which human flourishing depends. With the distributists, we recognize that character and community are formed through engagement with focal things and practices. Importantly, such communities welcome personal initiative and ingenuity, provided these energies are directed to creating productive tools, goods, and services that contribute to community well-being. Finally, distributists remind us of the need to find a proper balance between work and worship, action and contemplation. We share their acute awareness of modernity's hidden compulsions, manic pace, and incapacity to appreciate and make time for fruitful leisure and the *via contemplativa*.

How, then, might we take up the threads of distributist insight, in particular its wisdom regarding widespread ownership of productive property, and weave them into a new distributist program? The older distributist program was a "back to the land" plan, and we can endorse that option as a partial solution to our predicament. Toward this end, an expansion of land trusts and stronger land-reform laws that make possible a wider

---

27. Eric Gill, quoted in Lanz, *Beyond Capitalism & Socialism*, xx.

distribution of rural property are necessary, along with land-use policies that limit urban-suburban sprawl,[28] institutionalize best practices in sustainable agriculture, and ensure protection of biodiversity (e.g., through creation of biological corridors and buffer zones).[29] At the same time, a new distributist vision and program takes account of global demographic realities, i.e., the fact that most people will continue to reside in urban-suburban settings for the foreseeable future, while also defending a role for government that may make some distributists uneasy.

A new distributist politics of engagement and time will take many forms in these contexts, and here I suggest five priorities. First, restore the integrity and vitality of representative democracy by (1) instituting both public funding and limited private funding (e.g., the Tallahassee reforms discussed last chapter) of elections at all levels, and (2) expanding the use of instant run-off voting to encourage voter participation and a meaningful role for third parties, with experiments beginning at the local/county and state levels. Second, institute policies that favor local, cooperative production of goods and services such as food, clothing, furniture, health care, renewable energy, and education—all of which "lend themselves to engaging work and are capable of focal depth."[30] The federal government can provide crucial support for "indigenous innovation" (Phelps) and the creation of local banks and businesses embedded within and accountable to local communities by returning to an older tradition of strong antitrust legislation and enforcement as well as changes in intellectual property law that currently favor patent monopolies.[31]

Third, advance the practice of direct democracy by expanding and further refining participatory budgeting and inclusive planning processes at the local/metropolitan and bio-regional level. We desperately need institutional mechanisms that distribute political power more widely and foster greater involvement in public deliberation. In townhall meetings and other democratic venues, citizens shape their community's future, e.g., by taking control of energy production, funding high-quality public facilities that foster engagement with focal things and practices, protecting the commons,

---

28. For a Catholic social-ethical perspective on sprawl and land-use policy, see Nunez, "Ecological Common Good."

29. See Freyfogle, *New Agrarianism*. For a Christian-ethical critique of industrial agriculture in light of climate change and defense of a small-scale, sustainable agriculture alternative, see Graham, "Unsavory Gamble" and Northcott, *Place*, chap. 6. Within the church, a new distributist program supports the ongoing work of Catholic Rural Life.

30. Borgmann, *Technology*, 241.

31. Longman, "Economic Fates."

and building out a smart infrastructure under local-regional democratic control.[32] Fourth, put in place a carbon tax, with compensatory mechanisms for low-income consumers, to speed the transition to a post-carbon economy and provide (partial) funding for a basic income. And fifth, establish a basic income guarantee and reduced working hours at the national level (see chapter 8).

A new distributist politics also supports the creation of a humanistic information economy that respects producers, service providers, and users and promotes an ethos of fair and generous sharing.[33] As mentioned above, several promising initiatives for distributed social networking and digital platform cooperatives are under way. Encouraging examples of the latter include Fairmondo, a digital cooperative version of eBay started in Germany and run by seller-owners; Peerby, a Dutch neighbor-to-neighbor goods sharing platform with $2.2 million in crowd-sourced funding and plans to expand their "rent local from real people" model (Peerby Go) in the UK and North America; and Union Taxi, a driver-owned taxi cooperative based in Denver that offers an alternative to digital-monopolistic Uber.

To sum up, a new distributist politics will work toward the transformation over time of both urban-suburban and rural zones, guided by a broad vision of integral ecology and intent on building sustainable communities of varying sizes that achieve a "situated fitness" (Rolston) within ecosystems, not only bio-region by bio-region but also on macro-regional and continental scales.[34] New distributism prioritizes the expansion of a civil market economy governed by the logic of gift while seeking radical reforms of the

---

32. On public control of energy production, see Klein, *This Changes Everything*, chap. 3. One promising US example is the effort in Boulder, CO to reestablish municipal control over utilities.

33. Along these lines, more attention needs to be given to the groundbreaking work of Yochai Benkler and James Boyle, the latter a seminal voice in the cultural environmentalist school of thought among legal scholars and social scientists. Cultural environmentalism addresses issues of innovation, intellectual property rights, piracy, and the public domain. Just as environmental activists created a resonant discourse and symbolism around concern over the "endangered planet," cultural environmentalists are attempting to do the same with issues related to knowledge production, control of information, and the digital commons. The central question for this camp is: In the digital age, what cultural norms and legal regimes best encourage generation of new ideas and an optimal flow of information, knowledge, and culture? See Boyle, *Public Domain*.

34. Rolston's normative vision of sensitive earth residence attuned to a range of cultural and natural values discovered and conserved within three zones—urban-suburban, rural, and wild—is consistent with Francis's vision of integral ecology; see Rolston, *Conserving Natural Value*, chap. 1.

capitalist market, with the aim of putting the latter at the service of focal things and practices.

Granted, an appealing vision is one thing, a strategy for bringing it to fruition quite another. Hence in the next section I present a notable example of institutional change and market transformation before outlining a broad, three-pronged strategy for long-term systemic change in the US context.

A Notable Example of Institutional Change

The chapter's epigraph—a sweeping utopian statement announcing "the possibility of the regeneration of individuality, liberty, community, and ethics such as the world has never known, and a harmony with nature, with one another, and with the divine intelligence such as the world has never dreamed"—might be dismissed without comment had it not come from Dee Hock, the visionary founder and CEO emeritus of Visa, Inc., arguably the most successful business organization of all time. The creation of Visa presents us with a remarkable instance of how the adjacent possible moves from powerful ideas to reality, a case study from the business world that shows how innovation based on ecological principles can occur through a collective exercise in building shared vision and pursuing a noble purpose.

The Visa story begins in the mid-1960s, a time when a disorganized and chaotic credit-card industry was close to self-implosion. Hock was able to convince competitors within the banking world to cooperate for the sake of creating a national, and soon a global, system for the exchange of value based on principles operative in nature. In 1970, Visa, Inc. was born.

In his study of nature, Hock noticed that healthy, adaptive systems always exhibit a dynamic tension between chaos and order. Similarly, the design of Hock's "chaordic" Visa system allowed for a maximum degree of competition within an overarching context of cooperation. This design brought a fundamental tension into dynamic balance: Visa's member financial institutions issue the cards and compete fiercely with each other for customers, while at the same time they cooperate with each other—e.g., through rules-based participation in a common clearinghouse operation—to keep the system of value exchange running smoothly.[35]

As he observed and analyzed the modern business world, Hock became convinced that the hierarchical model of organization characteristic of the industrial revolution was deeply dysfunctional. Command-and-control organizations "were not only archaic and increasingly irrelevant. They were becoming a public menace, antithetical to the human spirit and destructive

---

35. Hock, *One From Many*.

of the biosphere. I was convinced we were on the brink of an epidemic of institutional failure."[36] The Visa organization—a non-stock, for-profit membership corporation with ownership in the form of non-transferable rights of participation—was highly decentralized and highly collaborative by design. It was exceptional in the way that authority, initiative, decision making, and wealth were pushed out to the members under the principle of subsidiarity.

Instead of trying to enforce cooperation by restricting what members can do, the Visa bylaws encourage them to compete and innovate as much as possible. "Members are free to create, price, market, and service their own products under the Visa name," Hock says. "At the same time, in a narrow band of activity essential to the success of the whole, they engage in the most intense cooperation."[37] This deft balancing of competition and cooperation enabled the system to expand worldwide despite the diversity of cultural environments and legal-political frameworks. "It was beyond the power of reason to design an organization to deal with such complexity," says Hock, "and beyond the reach of the imagination to perceive all the conditions it would encounter."[38] By incorporating principles of self-organization and organizational learning, Visa's system was able to thrive under complex, changing conditions. In Hock's view, "Visa is the archetype of the organization of the 21st century," one that mimics nature and respects the freedom and capabilities of all members within the organization. When Hock left Visa in 1984, it had overtaken Mastercard as the largest credit-card system in the world.[39]

Hock spent a decade in private life and then began to work with organizations interested in his unconventional approach to institutional design. He believes the key to reinventing organizational life is a people-centered process focused on the discovery of shared purpose and principles. The Visa system is no blueprint: "Far better than a precise plan is a clear sense of direction and compelling beliefs. And that lies within you. The question is, how do you evoke it?"[40] Building shared vision around purpose is a messy, time-consuming effort involving everyone in the organization, not simply senior leadership. At Visa, Hock practiced a form of servant leadership that

---

36. Dee Hock, quoted in Waldrop, "Trillion-Dollar Vision."
37. Ibid.
38. Ibid.
39. After Hock's departure, his unworthy successors reverted to more conventional management practices and reestablished a hierarchical structure within the organization.
40. Hock, quoted in Waldrop, "Trillion-Dollar Vision."

inspired others to take ownership of challenges and collaborate with co-workers in solving problems and innovating solutions.

According to Hock, after articulating an authentic and resonant statement of purpose, an organization then has to agree on a set of principles (e.g., subsidiarity, personal responsibility, mutual respect) that will govern the organization's structure, processes, and operations. These principles have to reflect what people in the organization believe in and really care about. While the process can be messy, frustrating, and time consuming, Hock argues it is absolutely essential because what people are trying to do is build a community. Only when the community has solid agreement on purposes and principles can it address the specific structure, processes, and operations of the organization. Not surprisingly, Hock's purpose-and-principles process for community building and organizational transformation has appealed primarily to civil-society organizations attuned to economic democracy.

Three Trajectories of Social Transformation

The Dee Hock/Visa story shows how significant change may occur when conditions allow a creative minority to advance concrete-utopian alternatives, e.g., a new institutional design or radical reform of public policy. On different scales, so do the stories from last chapter of Porto Alegre, with its shift to participatory budgeting as a way to overcome endemic corruption; Tallahassee, which replaced play-to-pay with strong ethics laws; Quebec, through an innovative social-economy model developed in response to endemic poverty and unemployment; Hawaii, by adopting distributed, renewable energy in place of expensive hydrocarbons; and Cleveland, where cooperative rather than corporate economic development combats urban decline. These examples fit into a broader strategic framework for advancing a new-distributist program and working toward the creation of a diverse, hybrid economy and transition to a socially inclusive post-jobs society.

Strategically, there are three trajectories of social transformation to pursue: displacement, radical reform, and disruption. While they require quite different forms of collective action at several scales and on occasion may work at cross-purposes, over time they tend to reinforce one another. Importantly, the envisioned process of social transformation does not involve a wholesale, revolutionary transition from capitalism to an entirely new social system. As Eric Olin Wright puts it, "emancipatory transformation should not be viewed mainly as a binary shift from one system to another, but rather as a shift in the configuration of the power relations that constitute a hybrid."⁴¹ Let us consider each strategic trajectory in turn, noting along the way how Catholic groups and institutions might contribute (and already do) to systemic change.

Displacement

The first transformational strategy involves a gradual displacement of corporate-neoliberal ideas, policies, and institutions from their hegemonic position through the patient, persistent building up of alternative institutions, which over time brings a new common good into being. A displacement strategy requires social innovation in a spirit of generosity and open sharing (think Wikipedia or The Freecycle Network) that puts the logic of gift into practice in creative ways. Alternatives come in many colors: producer and consumer cooperatives foremost, but also co-housing settlements, worker-owned companies, neighborhood corporations, land trusts, CSAs and farmers' markets, bartering and timeshare groups, social enterprises, B corporations, local currencies, and various community-based groups focused on inclusive, grassroots economic development. As these groups and organizations connect through networks and various forms of cooperative association, they mimic nature by creating the socioeconomic counterpart to a rhizomatic web, i.e., a resilient and complex root system spreading laterally and thriving amidst more dominant root structures (e.g., a big-box store, a retail shopping mall, a public housing complex). When this occurs within local economies, opportunities expand for workers and consumers to opt out of the corporate-dominated formal economy through participation in a "local, living economy."⁴²

Consistent with Catholic social tradition's emphasis on subsidiarity, pluralism, and dignified social participation, a displacement strategy focuses on long-term community wealth-building and democratic renewal from the bottom up. Ingenuity and capability-building are encouraged in this approach. A good US Catholic example is found in the previously discussed Appalachian bishops' pastoral message "At Home in the Web of Life," which sets forth an alternative vision of sustainable community. On the ground, examples of cooperative, not-for-profit, and sharing models successfully at work on a small scale include the Catholic-sponsored Solidarity HealthShare, fair-trade networks, and various grassroots projects supported

41. Wright, *Envisioning Real Utopias*, 367.

42. For a defense of this strategic approach against arguments from the left that it reinforces rather than challenges neoliberal hegemony, see Schor and Thompson, *Sustainable Lifestyles*, chap. 8.

by the Catholic Campaign for Human Development. Benedict's "economy of communion" currently exists in faith-based forms (e.g., Focolare) and in the wider society (e.g., the Mondragon cooperatives, which were founded by a priest inspired by Catholic social teaching).

Radical Reform

The second strategy seeks to advance a new distributist policy agenda involving radical reform at every level of governance, from local ordinances to global treaties. While priorities and action strategies will vary across geographies, the general thrust involves a radically reformed legal-regulative framework supporting the transition to a more diverse, hybrid economy oriented to long-term performance, resilience, and stability rather than GDP growth and short-term profits. Earlier, I mentioned reforms of land ownership and use to encourage a revival of rural communities; a shift to policies that favor local, cooperative production of goods and services; expanding practices of participatory budgeting, community-based planning, and other forms of direct democracy; establishing a carbon tax to accelerate the shift to a post-carbon economy; and instituting a basic income and reduced work week. To succeed, such reforms will require a dismantling of neoliberal trade policies and WTO rules that undermine democratic control over local-regional economic development and grant corporations free rein to play localities and regions off against one another. Other major reforms include state and federal laws to ensure democratic control of the financial sector, clean government, and strong environmental standards.[43] At its core, this policy agenda promotes engagement with focal things and practices, giving priority to quality of life and social inclusion over affluence. In the US, statewide Catholic advocacy groups, the national bishops' conference, and the PICO National Network of faith-based community organizations are among the avenues for advancing elements of a new distributist policy agenda.

43. Admittedly, this wish list begs a number of questions. In the US context, one is the issue of widespread political disengagement and sense of disempowerment, another the constraints of a two-party system and two-year election cycle, a third the distressing drift into polarization and mistrust within the political culture, a fourth the widening social distance between people from different socioeconomic backgrounds—all of which makes it difficult indeed to advance any alternative political agenda. That said, a long-term strategic outlook and concrete-utopian imagination looks for opportunities to advance radical reforms. Crazy ideas dismissed today have a way of becoming tomorrow's common sense. For discussion and case studies of innovation in deliberative-democratic practices and institutions, see Smith, *Democratic Innovations*.

Disruption

The third strategy involves disruption of business as usual using nonviolent direct-action methods to defend the vulnerable, slow down or halt damaging corporate-neoliberal projects, and generate the "creative tension" (King) necessary for moving policy reforms forward. If and when a politics of radical reform makes progress, then it is sure to face aggressive push-back from threatened capitalist elites. Bipartisan efforts will be sidelined; appeals to dialogue and collaboration will be met with intransigence from entrenched interests. Thus countervailing power in the form of creative nonviolence will be necessary to prevent further damage, alert the public to what is at stake, delegitimize efforts to roll back reforms or perpetuate unsustainable practices, and create a political climate more conducive to systemic change. To paraphrase Reinhold Niebuhr's famous observation, humanity's capacity for justice makes democracy possible, but the inclination to injustice among the powerful makes nonviolent direct action in defense of democratic values and planetary life systems necessary.

Perhaps the best example of disruptive nonviolence in the US currently is the movement in North Dakota and elsewhere to prevent the hydrocarbon complex from locking in infrastructure projects while calling attention to the need for a major policy commitment to a post-carbon, post-nuclear energy future. This movement represents a withdrawal of consent by the governed and direct challenge to rentier interests hell-bent on expanding an unsustainable infrastructure of energy production and consumption.[44] Another good example is Rolling Jubilee, a grassroots effort to relieve lower- and struggling middle-income households of student, medical, and credit card debt.[45] Those in solidarity with frontline activists take direct action by modifying patterns of consumption, experimenting with new role

---

44. The stakes here could not be higher, as pointed out by Bill McKibben, Naomi Klein, and other climate-justice activists. To keep within the IPCC benchmark of a two degrees Celsius rise in global mean temperature in the current century, approximately 80 percent of known reserves of fossil fuel must remain in the ground, i.e., a staggering $10 trillion in fossil fuel wealth must be given up by its owners as the global economy transitions to post-carbon energy systems. The only historical parallel is the loss of "property" by Southern slave owners; see Hayes, "New Abolitionism." However, as noted above, oil may become the feedstock for advanced construction materials that scale, thus allowing Big Oil to survive the end of the internal-combustion engine and contribute otherwise stranded assets to the quest for net zero construction; see Doherty et al., "Oil and Gas."

45. On the uses of debt as an instrument of domination and control by elites throughout history, see Graeber, *Debt*. An anthropologist and leading Occupy activist, Graeber's work inspired the Rolling Jubilee movement (www.rollingjubilee.org).

performances,[46] and forging new lifestyles and intentional communities oriented around focal things and practices. Catholics and other Christians called to faith-based nonviolent witness today find inspiration and practical lessons in Dorothy Day and the Catholic Worker movement, the Berrigan brothers and Ploughshares peace activism, and School of the Americas protests.

To sum up, a "flexible strategic pluralism" (Wright) calls for (1) long-term community wealth-building and democratic renewal from the bottom up; (2) radical reforms advanced at every level of government; and (3) nonviolent acts of disruption to protect the vulnerable, delegitimize the status quo, and generate creative tension. Committed to a politics of engagement and time, Christians do well to join with others in pursuing all three transformational strategies, with some dedicated to distinctly Catholic and ecumenical projects and others involved in social movements, civil-society organizations, and activist networks operating in the wider society. What may be possible is the creation of a diverse, hybrid economy and more socially inclusive post-jobs society geared toward sustainable abundance for all rather than unsustainable affluence for the few.

Dark Possibilities and Christian Praxis

We cannot know, of course, whether emerging conditions will favor or make more difficult efforts to execute on a long-term, transformational strategy. Doubtless a rethinking and reframing of the adjacent possible will be required as the future unfolds. That said, we can point to growing evidence suggesting that systemic contradictions and accelerative compulsions are likely to generate shocks and crises of increasing scale and magnitude. Hence it is not idle speculation to ask how individuals, organizations, communities, and nations may respond if the respective visions of neoliberals, technocrats, and comfortable moderates all prove illusory. Will a modern-day progressive movement arise to save global capitalism from itself? If not, and growing numbers lose faith in a faltering system, will they seek out alternatives that move beyond piecemeal reform or lapse into resignation and despair? Will those at the commanding heights resent calls for radical reform and adhere to received scripts of economic growth and affluence, or will some adopt new roles as transformational leaders (or useful transitional

---

46. On how various forms of nonviolent disruption through role experimentation in many different micro-political environments may contribute to broader and more militant political action, see Connelly, *Fragility of Things*, 179–80.

figures)? How many among the religious will lose faith or take refuge in apocalyptic while acquiescing to business as usual?

Again, we can't know in advance how severe the problems of global debt, poverty, inequality, environmental degradation, and rising structural unemployment may (or may not) become in the decades ahead, nor can we predict how individuals and institutions will respond if these systemic risks do reach critical thresholds and morph, perhaps quickly, into full-blown and mutually reinforcing crises that make social transformation on a broad scale absolutely necessary for a decent global society to exist at all. If tipping points are reached and severe or worst-case scenarios do unfold rapidly, then how capable is humanity as a whole—and how capable are particular communities and societies—of transforming attitudes, lifestyles, social practices, and institutions for the sake of sustainable abundance (optimal case) or simply for survival in a radically altered and very harsh environment (minimal case)? Under severe crisis conditions, a descent into social chaos followed by regression to authoritarian rule or barbarism cannot be ruled out.

It would be unwise to dismiss such dark and seemingly remote possibilities, if only because a small but growing chorus of sober, well-informed voices now argue that a "Great Disruption" is coming whether we are prepared for it or not. Optimists such as Paul Gilding argue that a wave of severe global crises—e.g., more frequent and prolonged shortages of food, water, and fuel in many countries along with increasingly deadly extreme weather events—will break through the heavy dam of collective denial and release immense waves of human energy and creativity, and hence broad social change will occur amidst our darkest hour, but not until then.[47] Pessimists such as Hartmut Rosa counter that modernity's social-acceleration processes possess too much power and momentum, and hence ecological overshoot and collapse are the most likely prospect. James Lovelock agrees: when the full extent of destabilizing anthropogenic changes to Gaia's complex, interrelated systems become manifest, most human settlements won't be able to manage a tumultuous transition to the new, post-Holocene normal. Lovelock's worst-case scenario has a few hundred million survivors of the seismic shift to a "new equilibrium" resettling in select geographies still capable of food production.[48]

From a faith perspective, we can see already how much human suffering and wasted potential goes unaddressed and acknowledge as current fact that mass species extinctions are happening on our watch, and hence we

---

47. Gilding, *Great Disruption*.
48. Lovelock, *Vanishing Face of Gaia*.

can direct our energy and creativity toward pursuing the adjacent possible of sustainable abundance now. Amidst irreducible uncertainty, we must act nonetheless or risk ceding the field to those who cling to familiar narratives and counsel more of the same, or who argue that it's already too late to avoid eco-apocalypse and succumb to resignation and despair. Liberating faith and the shared exercise of a prophetic-utopian imagination by Christian communities can awaken hope and energize collective action informed by long-term strategic vision even in the midst of difficult circumstances and a growing sense that we are indeed in a moment near midnight.

Seeing the Whole

Going forward, the renewal and honing of democratic energy and cultural creativity is a *sine qua non* for moving toward a more just, humane, and sustainable world. As the stories from this chapter (Visa) and the last suggest, social innovation can thrive through distributed power, self-organization, and the continued harnessing of info-tech for high productivity. Wikipedia provides another example, Stanford's Folding@home project still another. Those who possess such energy and creativity are the ones outside Saba, as it were, who can envision the adjacent possible in a spirit of hope, and who can live and act in ways that foster both personal and social transformation, even though they have no corner on the future. A new-distributist program and three-pronged transformational strategy will be advanced by visionary leaders such as Dee Hock who are capable of "seeing from the whole," i.e., of sensing the formative field in which patterns (or waves) unfold and all things evolve. "When we become more aware of the dynamic whole, we also become more aware of what is emerging."[49] As Bloch says, the real-possible comes into view through a marriage of reason and imagination. Where hope and courage live, both selves and situations can become otherwise. A dynamic, unfolding plenitude continually presents to attentive, empathetic, and imaginative leaders new opportunities for the convivial co-creation of greater value and beauty on the home planet.

---

49. Senge, *Presence*, 12.

8

Good Lives, Automation, and the Transition to Post-Jobs Society

Eight hours for work, eight hours for rest, and eight hours for what we will.
—LABOR SLOGAN FROM LATE-NINETEENTH CENTURY

It is not that men are ill fed, but that they have no pleasure in the work by which they make their bread.
—JOHN RUSKIN

Automation confronts us with the most important question of all: What does *human being* mean?
—NICHOLAS CARR

Goods We Pursue in Common

In today's world risk society, socially conscious Catholics and other Christians do well to think holistically and systemically about environmental degradation, poverty, rising inequality, polarized politics, and another inconvenient truth on the horizon: accelerating automation and the specter of rising structural unemployment. Prophetic-utopian thinking devoted to connecting the dots between these daunting challenges and discerning prospects for social change might begin by asking two distinct yet complementary questions: What constitutes the good life, and how can human and planetary flourishing be realized in a risky, runaway world? Where do opportunities for pursuing the adjacent possible appear most promising within the current, complex situation?

Reflecting on the eco-social question can help us to envision and pursue the adjacent possible of sustainable abundance for all by providing a normative social vision of the good life and human/planetary flourishing, and by stimulating dialogue and collaboration around strategies for moving collectively "from here to there." Whether it is named a civilization of love (Paul VI), integral ecology (Francis), genuine wealth (Borgmann), or "sustainable prosperity for all,"[1] Catholic social thought presents us with a teleological view of the good life and a holistic vision of the good society grounded in a theological anthropology. To recall the discussion from chapter 6, we humans flourish and find true fulfillment when our "fundamental option" (Rahner) and basic stance in life is oriented to the transcendent dimension of existence, i.e., is fundamentally *open* to a deeper and more intimate relationship with God, who is our absolute future. Each of us is an evolving body-brain-heart nexus—an embodied soul—impelled toward greater knowing and loving in freedom and in communion with others. We grow and change through acts of self-transcendence within the context of healthy, supportive communities. At widest scope, our vocation and comprehensive purpose is the convivial co-creation of greater value and beauty amidst unfolding plenitude in universal history.

On the home planet, that vocation becomes concrete and dramatic (as in artful performance) in and through the various roles we play and narrative identities we form within communities of focal practice. In plain language, what we really want and find fulfilling over a lifetime are six *basic goods*: physical and emotional health, opportunities for learning and personal growth, close relationships, meaningful work (paid and unpaid), fruitful leisure (which includes socializing, deep play, travel, and prayer/worship), and higher purpose (i.e., belonging to and serving something greater than ourselves, with loving devotion to God the ultimate *raison d'être*). In fabulously diverse ways, these are the basic goods we pursue in common, and importantly they represent a common ground, as it were, for dialogue and collaboration aimed at moving toward a world of sustainable abundance for all.

1. This phrase comes from the True Wealth of Nations project, which gathers together Catholic scholars from various fields to investigate how the church's social teaching addresses economic life. The group believes, as I do, that Catholic social wisdom offers a much needed alternative to neoliberal-capitalist and state-socialist models.

All of these basic goods are intrinsically rewarding and contribute to good lives, as a growing body of positive psychology research attests. For their realization on a regular basis by society's members, they require sustained commitment both to personal development (or excellence) and to the common good. Moreover, a good society will protect and promote a range of focal things and practices that enable persons to pursue basic goods in the context of healthy, supportive communities. Finally, a good society is structured by policies and institutions that ensure real freedom, security, and opportunity for all as preconditions for pursuing lives well lived.

The previous chapter discussed the systemic risks generated by global debt, climate change, poverty, and growing inequality while alluding to the emerging issue of technological unemployment. In this chapter I argue that exponential technologies are likely to accelerate automation and thus eliminate tens of millions of jobs across the globe in coming decades. The ensuing jobs crisis, already acute in many countries today, needs to be met with a new strategy that simultaneously addresses these other major concerns. As structural-unemployment levels rise, socially engaged Christians are called to articulate and act on a hope-filled vision of sustainable abundance for all and prepare the way for a *post-jobs society* in which today's norm of an eight-hour work day and forty-hour work week gives way to less time on the job and more time for what we will. A politics of engagement and time will view the public's growing awareness of accelerating automation as an opportunity to advance a new distributist policy agenda centered on a basic income guarantee (BIG) and thirty-hour work week. At stake are the enabling conditions for pursuing the basic goods that constitute human flourishing.

I first consider the impact of emerging technologies and secular trend of technological unemployment. I then argue for making automation a central issue in public discourse and linking it to concerns about poverty, inequality, and ecological limits to growth in ways that are empirically credible and that open hearts and minds to consider alternative horizons. A third section sketches out how a just, humane transition to a socially inclusive post-jobs society might occur and offers a Catholic defense of basic income and reduced work week. The fourth section examines more closely the idea and ideal of post-jobs society, i.e., a society in which most people work fewer hours for pay and are thus free to devote more time to family life, community service, deep play, contemplation, and other worthy pursuits. A final section explores what I call the "priority of the cultural," where Catholics and other persons of faith commit to leading the way toward redefining the good life and creating a post-jobs society in which *meaningful work* (paid and unpaid), rather than full employment and GDP growth, is the policy priority and various forms of fruitful leisure are honored and enjoyed more fully than at present.

The Second Machine Age

In an essay titled "Economic Possibilities for our Grandchildren" (1930), John Maynard Keynes wrote: "We are being afflicted with a new disease of which some readers may not yet have heard the name, but of which they will hear a great deal in the years to come—namely, technological unemployment. This means unemployment due to our discovery of means of economizing the use of labor outrunning the pace at which we can find new uses for labor."[2] Paul Krugman, Robert Reich, and other prominent policy experts now argue the risk of technological unemployment is rising.[3] Notably, MIT researchers Erik Brynjolfsson and Andrew McAfee take issue with the received view, which holds that new technologies create only short-term disruptions for workers. While innovation displaces labor, it also spurs demand for the cheaper, better products coming out while introducing new inventions that reshape economic practices and create new types of jobs. Rising demand stimulates production and thus greater demand for more and different kinds of labor as the economy diversifies and expands, increasing the net total of job categories and employed workers. Against this prevailing technology-generates-jobs theory, Brynjolfsson and McAfee point to mounting evidence from 2000 onward of a "great uncoupling" in which productivity gains from automation are de-linked from employment growth for the first time in modern history.[4]

According to Brynjolfsson and McAfee, both the great uncoupling and rising inequality are worrisome aspects of what, on balance, is an enormously promising transition to what they call the "second machine age." During the first machine age—powered initially by steam and then by electricity and hydro-carbons—new industries generated steady gains in productivity, living standards improved, life spans lengthened, and populations grew. While automation reduced demand for farm and factory labor, a growing

---

2. John Maynard Keynes, quoted in Frey and Osborne, "The Future of Employment."

3. Krugman, "Sympathy for Luddites"; Reich, *Saving Capitalism,* chap. 23.

4. Brynjolfsson and McAfee, *Second Machine Age.* Recently a group of AI researchers, engineers, and scientists echoed Brynjolfsson and McAfee's concerns in an open letter titled "Research Priorities for Robust and Beneficial Artificial Intelligence" on the Oxford-based Future of Life Institute's website. The letter's fast-growing list of signatories, including Stephen Hawking and Elon Musk, identified a set of research priorities that included policy responses to AI's adverse impact on labor markets.

service economy absorbed displaced workers. Up to 1980, education and training levels kept pace with technology; workers were able to meet job requirements across a range of industries.[5] Throughout the twentieth century, the disease of technological unemployment was kept at bay.[6]

As the second machine age gathers force in coming decade, Brynjolfsson and McAfee argue that productivity gains from exponential technologies will dwarf those of the previous century. Most importantly, Moore's Law—the doubling of computer power every two years—is spreading to other technologies. Global society has reached an inflection point, a historical moment when digitalization, machine learning, and other innovations will spur a productivity revolution.[7] They note how "the exponential, digital, and recombinant powers of the second machine age have made it possible for humanity to create two of the most important one-time events in our history: the emergence of real, useful artificial intelligence and the connection of most of the people of the planet via a common digital network."[8] Possibilities for recombinant innovation are growing rapidly; as new ideas and inventions multiply, myriad other options for novel technology combinations open up. The IoT exemplifies this trend. Currently we see a combination of computer networks, sensors, actuators, sophisticated analytics, and advanced algorithms harnessing Big Data and bringing the IoT to life across the globe. A McKinsey Global Institute study from 2013 forecasts the IoT will generate dramatic gains in efficiency and productivity within the manufacturing, health care, and mining industries over the next twelve years.[9] As noted last chapter, Jeremy Rifkin and others see a wider and deeper transformation to come.

What, then, is the fate of formal employment in the second machine age? The technology-jobs equation will undergo major changes as automation accelerates and the source of wealth shifts to idea generation and control over information flows. On the upside, more opportunities for non-routine, cognitive work will open up. The well-educated, creative, technically skilled,

---

5. Goldin and Katz, *Race between Education and Technology*.

6. For a historical overview of technology's impact on economy and employment, see Mokyr et al., "History of Technological Anxiety."

7. Jan Vijg, Robert Gordon, and Tyler Cowen hold a diametrically opposite view, arguing that flagging technological development translates to stagnant productivity and economic stagnation in the decades ahead.

8. Brynjolfsson and McAfee, *Second Machine Age*, 90.

9. The McKinsey study, titled "Disruptive technologies," identifies a dozen cutting-edge technologies that are likely to achieve high-impact status by 2025. For example, advanced robotics has the potential to affect $6.3 trillion in labor costs globally over the next decade

and entrepreneurial will thrive along with superstar performers in select fields such as the arts, professional sports, media, and entertainment. As Brynjolfsson and McAfee note,

> Machines are substituting for more types of human labor than ever before. As they replicate themselves, they are also creating more capital. This means that the real winners of the future will not be the providers of cheap labor or the owners of ordinary capital, both of whom will be increasingly squeezed by automation. Fortune will instead favor a third group: those who can innovate and create new products, services, and business models.[10]

The build-out of an IoT-anchored smart infrastructure also will stimulate employment on a large scale. Advances in solar technology and battery storage now make renewable energy competitive with fossil fuel and nuclear options. Globally, these developments promise to generate millions of green jobs as renewable energy grids are constructed, buildings are retrofitted and designed to become micro-power plants, and a new transportation system friendly to electric and fuel-cell vehicles is laid out.

However, despite these projected growth areas in employment, we can anticipate that unemployment, underemployment, and contingent labor will increase as exponential technologies scale up and businesses opt for automation while continuing to push hard for flexible employment arrangements. With markets globalized, intensification of the winner-take-all effect will contribute to greater inequality as well as more severe labor market polarization.[11] Even presuming gains in access to training and education, the skills gap likely will widen between what many ordinary workers can offer and what employers demand from them, both in developed and developing countries.[12] Tomorrow's global economy will tend to generate more lovely and lousy jobs, but is less likely to generate more decent-paying, middling jobs that smart machines can do more efficiently. Absent major policy interventions, economists expect both inequality and unemployment levels to increase during the second machine age.[13]

As the global labor force grows to an estimated 3.5 billion in 2030, accelerating automation will eliminate many hundreds of twentieth-century job categories across dozens of industries and throw tens of millions out of

---

10. Brynjolfsson et al., "Labor, Capital, and Ideas," 47.
11. Brynjolfsson and McAfee, *Second Machine Age*, chaps. 9–11.
12. On global labor force trends through 2030, see McKinsey, "The world at work."
13. Tyson and Spence, "Effects of Technology on Inequality."

work.¹⁴ Economists already see a "premature de-industrialization" (Rodrik) taking place in India and other global South countries due to automation. Moreover, these projected trends will unfold at a time when, according to Gallup, there already exists a 1.8 billion shortfall in "good jobs" (i.e., decent-paying and reasonably secure full-time employment) across the globe.¹⁵ And among those working, a growing number will be in lower-paying and less-secure contingent roles as more companies shift toward atomized, networked labor processes and just-in-time production models. Over 40 percent of US workers are now part-timers, contractors, temps, on-call, or self-employed in the gig economy. By 2020 it is likely that over half the American workforce will be in the contingent category.

The fate of less-skilled, less-educated workers in this scenario is troubling enough, but with smart-technology development taking off many white-collar jobs also will be partly or entirely replaced. A widely cited 2013 study of future US employment by Carl B. Frey and Michael A. Osborne at Oxford indicates that low-wage workers performing routine tasks are most susceptible to computerization. As of 2014, the four most-common occupations in the United States—retail salesperson, cashier, food and beverage server, and office clerk—employed a combined 15.4 million people, close to 10 percent of the workforce. Workers in these four occupations are highly vulnerable to automation. The research also suggests that advances in smart-machine design will allow companies in many industries to automate a number of job types characterized by non-routine, manual labor as well as many types of non-routine, cognitive work (e.g., legal researcher, copy editor). Until recently, these non-routine jobs generally were considered immune from automation, but the cost-benefit calculus is being altered by advances in mobile robotics, machine learning, sensor technology, and other fields.¹⁶ Frey and Osborne conclude that *47 percent* of total US employment is at "high risk" of being eliminated by automation over the next decade or two, but that startling figure could be even higher if one includes

---

14. See Ford, *Lights in Tunnel*, chaps. 2–3. Economist W. Brian Arthur sums up where we've been and where we're at now: "For centuries, wealth has traditionally been apportioned in the West through jobs, and jobs have always been forthcoming. When farm jobs disappeared, we still had manufacturing jobs, and when these disappeared we migrated to service jobs. With this digital transformation, this last repository of jobs is shrinking—fewer of us in the future may have white-collar business process jobs—and we face a problem" (Arthur, "Second Economy").

15. Clifton, *Coming Jobs War*, 2.

16. On how advanced technologies are encroaching on white-collar jobs, see Ford, *Rise of Robots*, chap. 4.

some percentage of non-routine job categories that they put at "medium risk."[17]

Chinese companies are moving aggressively to automate production. Take the well-known case of Foxconn as an example. The world's largest contract manufacturer of consumer electronics,[18] Taiwan-based Foxconn employs over a million workers, most of whom are low-paid ($1.50–$2.20 USD per hour) Chinese on the mainland. Over the past decade, actual and threatened worker suicides at Foxconn's huge facilities have garnered public attention. Labor rights activists have found poor working conditions, patterns of mistreatment, and payment of subsistence wages. In response, founder and CEO Terry Gou announced in 2011 a major new plan to replace most of Foxconn's massive workforce with one million robots. While the first wave of Gou's "Foxbots" reportedly have underperformed, industry analysts fully expect the technology to improve in coming years, at which point automation at Foxconn (and elsewhere) will scale up.[19] While the company anticipates large cost savings, beleaguered employees wait for pink slips.

Jaron Lanier, whose proposal for democratizing cyber-wealth I discussed in chapter 7, argues that rising inequality and technological unemployment are linked to a flawed network design.[20] The current digital architecture allows elites to build and control what Lanier calls "Siren Servers": a "powerful computational resource that out-competes everyone else on the network and seems to grant its owners a guaranteed path to unbounded success at first."[21] With their immense computing power, sophisticated analytics, and advanced algorithms, Siren Servers enable their owners to master Big Data and attain dominant market positions. By exploiting extreme forms of information asymmetry, Siren Server organizations such as Google, Facebook, Amazon, and Walmart are able to dissect systematically the behaviors of competitors, suppliers, and customers. The result is less risk for them and more risk for everyone else.[22]

Over time, Siren Servers undermine the economy by reducing employment and extracting value from localities. As Lanier puts it, "People are created equal, but computers are not. A top computer can bring limitless

17. Frey and Osborne, "Future of Employment."
18. Foxconn manufactures well-known brands such as BlackBerry, iPad, iPhone, Kindle, PlayStation 4, Xbox One, and Wii U.
19. Brunner, "Foxconn replaces human workers."
20. Lanier, *Who Owns the Future?*
21. Ibid., xxiii.
22. Cf. Rushkoff, *Throwing Rocks at Google*, 82–93.

wealth and influence to that lucky computer's owner and the onset of insecurity, austerity, and unemployment for everyone else."[23] He cites the fate of Kodak as a case in point. At its height fifteen years ago, Kodak employed more than 140,000 and had a net worth of $28 billion. Today, the company that invented the digital camera is bankrupt, and the future of digital photography belongs to the likes of Instagram. In 2012, when Facebook bought Instagram for $1 billion, the startup had just thirteen employees. Given the present design of digital networks and dominating role of Siren Servers, Lanier sees one industry after another going the way of music, journalism, and photography. As computer networks become more powerful, the IoT takes hold, and smart machines make investment in automation increasingly attractive, unemployment and underemployment levels will rise.

Left unaddressed, rising inequality and technological unemployment will take a heavy toll on both individuals and society. In *The Spirit Level*, public-health researchers Richard Wilkinson and Kate Pickett present abundant empirical evidence detailing the negative social effects and economic costs of inequality in countries with a high Gini coefficient (a measure of inequality).[24] High unemployment and underemployment rates for young workers threaten to become a permanent feature of the social landscape in both developed and developing countries. Millennials struggling to begin a career, and often chronically underemployed, face a higher likelihood of reduced earnings, which means less savings and hence less investment in the future.[25] Psychologically, chronic underemployment and prolonged unemployment corrodes a sense of self-worth. In a society that equates power, security, and status with formal employment, many under- and unemployed persons lack the resources and support to reconstitute their identities through other roles and activities, and those who "drop out" typically are stigmatized.

From a societal perspective, the negative consequences of high unemployment are well known. The long-term unemployed become the socially excluded and marginalized. In communities with high unemployment, all the indicators of social well-being move in the wrong direction. Rising unemployment translates into higher rates of crime, domestic abuse, divorce, depression, suicide, and violent unrest.[26] In turn, a law-and-order response—institutionalized through formation of a prison-industrial

---

23. Lanier, *Who Owns the Future?*, xxv.
24. Wilkinson and Pickett, *Spirit Level*.
25. Sachs and Kotlikoff, "Smart Machines and Long-Term Misery."
26. Wilson, *When Work Disappears*.

complex—succeeds only in creating a vicious cycle of reduced opportunity, hopelessness, and deeper alienation, particularly for people of color.[27]

Time to Reframe

With the official US unemployment rate trending downward in recent years, the specter of technological unemployment has yet to find its way into the mainstream, despite growing attention to the issue in the media and blogosphere. Historically, workers feared displacement from mechanization during the Depression, and with first-wave cybernation the issue resurfaced in the postwar decades. A guaranteed income bill (Nixon's Family Assistance Plan), prompted by concern over poverty and technological unemployment, passed the House in 1972 but failed in the Senate.[28] With the shift to post-Fordist production and waves of downsizing impacting white- and blue-collar workers alike, sociologists took up the issue during the 1990s.[29] These political and academic debates revolved around two basic questions: Who owns and controls the means of production, and how are technology-driven productivity gains to be shared? Such questions will return with renewed force as the downside of the second machine age comes more clearly into view and public awareness of persistent poverty, more pronounced inequality levels (as forecasted by Piketty), and the disruptive impacts of climate change grows.

For their part, church leaders have yet to address fully the ethical and political questions raised by exponential technology and accelerating automation. On matters of work and economic justice, the focus continues to be on workers' rights, calls for full employment, and efforts to address the injustices of domestic and global inequality (e.g., the living wage, fair trade, and Jubilee movements). Socially engaged Catholics and other religious progressives have articulated a sharp critique of neoliberal globalization and set forth more humane, just, and democratic alternatives. These alternative social visions largely have been ignored by the American public, yet many yearn for a credible alternative to the frantic pace of life and heavy debt burdens they experience under predatory capitalism, on the one hand, and the heavy hand of government interventions they perceive as clumsy and corrupt, on the other. Wary of turning either right or left, ordinary citizens struggle to see how the country can move forward.

27. Sudbury, *Global Lockdown*.

28. Bix, *Inventing Ourselves Out of Jobs?*

29. Aronowitz and Cutler, *Post-Work*; Gorz, *Reclaiming Work*; Rifkin, *End of Work*; Beck, *Brave New World of Work*.

I suggest the time for reframing the situation has arrived, based partly on what we know about current public attitudes toward class and redistribution and partly on foresight into how certain trends likely will play out in coming years. We know that growing middle class insecurity stems from decades of wage stagnation, jobless recoveries, and high household debt levels.[30] Significantly, nearly ninety percent of US adults self-identify as "middle class."[31] The stress and anxiety felt by many (self-identifying) members of the middle class presents an opportunity to reframe the situation and offer a desirable, viable alternative to the current political deadlock. A politics of engagement and time seeks to do just that by showing how two powerful policy ideas, namely basic income and a thirty-hour work week, can address in tandem the issues of poverty, inequality, automation, and climate change.

Below I elaborate on the ways in which a basic income guarantee (BIG) and reduced work week can transform our moribund politics, promote healthy changes in cultural norms related to work and leisure, and offer genuinely effective remedies to the four issues I've flagged. At this point, it may prove helpful to indicate briefly how a BIG in particular can contribute to meeting these four challenges. First, as labor leader Andy Stern and others now contend (and as Dr. King argued a half-century ago), providing a basic income to all US adults will eradicate persistent poverty; tens of millions of poor Americans, an increasing number of whom now live in extreme poverty (i.e., barely surviving on two dollars a day or less),[32] will find their life prospects dramatically improved, while society will benefit enormously from reduced health and welfare costs and other economic gains related to the elimination of poverty. Empowering the poor is a Christian-ethical imperative, and so it is no small matter that recent research supports the argument for liberating them from fear of want, and the daily toll it takes on their psyche, through stable provision of a basic income.[33]

Second, over time the redistributive effect of a BIG will be to reduce inequalities in wealth and income somewhat (estimates vary). As Wilkinson and Pickett show, greater socioeconomic equality contributes to a stronger economy and happier populace in good part because society's lower-income members are more healthy and productive. In this way, a BIG yields a win-win for rich and poor alike. Third, combined with a thirty-hour work week,

30. Cohen, "Middle Class."

31. Pew, "Government Policies."

32. Edin and Shaefer, *$2.00 a Day*. The researchers find that 1.5 million US households with over 3 million children now subsist on $60 or less per month.

33. On the evidence-based, economic case for ending poverty, see Bregman, *Utopia for Realists*, chap. 3.

a BIG will mitigate the negative impact of automation in several ways, e.g., by giving workers the means to retrain, accept apprenticeships, work part-time, volunteer, or take a lower-paying job they are passionate about.

Fourth and finally, any serious attempt to mitigate climate change *and* address other pressing environmental concerns such as loss of critical habitat will require deep de-carbonization, redesign of built landscapes to achieve a situated fitness within watersheds and ecosystems, and lower aggregate levels of consumption. Both a BIG and a thirty-hour work week will contribute significantly toward meeting the goals of a sustainable society as more people are able to adopt post-scarcity, non-material lifestyles and take up paid and unpaid work involving either a reduced ecological footprint (e.g., caretaking, teaching) or regeneration of local ecologies. While the human health-and-happiness effects of a shift to dynamic, steady-state economies will be considerable, the other major rationale for this transition is *planetary flourishing* through protection of the many wild and rural areas needed to sustain biodiversity. In short, the BIG idea provides a powerful, if partial, answer to the eco-social question. It will help to provide the enabling conditions for pursuing good lives and protecting the planet.

By contrast, a singular focus by the left on growing inequality and redistribution faces an uphill battle in the current political environment. While counterintuitive to Sanders supporters and other progressives, empirical studies show that rising inequality in the US has not translated into more public support for redistributive policies. In fact, just the opposite holds: rising inequality is correlated with ideological conservatism among higher- *and* lower-income groups on the question of redistribution to the poor and unemployed in a context of increased fear of ethnic-religious diversity from immigration, widening social distance, growing distrust of government, and political polarization. The stoking of nativist and racist sentiment by right wing populists has revived the pseudo-problem of the "undeserving poor" (i.e., working-class immigrants and inner-city residents), which continues to distort public perceptions of poverty and welfare policy. Merely tepid support for redistribution from the rich to the rest of society has held steady among middle- and lower-income groups, despite the growing gap in income and wealth between the one percent and everyone else.[34]

As frustrating as it may be for social justice activists, growing inequality is not a problem the American public is exercised about enough to take serious collective action on (at least not yet), the Occupy movement, recent

---

34. See Luttig, "Structure of Inequality"; and Ashok et al., "Support for Redistribution." In addition to confirming the no-increase finding of other research, Ashok et al. find that support for redistribution has decreased among the elderly and African-Americans in recent years.

gains by living wage campaigns, and activation of social-democratic forces during the Sanders presidential run notwithstanding. Nor has growing public concern over climate change translated into a major policy initiative at the national level. Instead, the Trump administration is busy rolling back hard-won environmental policies, including repudiation of the Paris climate agreement, while ignoring veteran Republican leaders who have proposed a carbon tax and dividend plan that would address climate change in a serious way and introduce to citizens the idea of, you guessed it, a basic income (through what they call a "carbon dividend").[35]

Meanwhile, awareness of automation's far reaching impact on the workplace is growing, albeit slowly at present. However, as unemployment and underemployment levels rise with the next economic downturn, socioeconomic inequalities become even more glaring, and the consequences of inaction on climate change become more obvious and worrisome, we can expect public concern to reach a tipping point and calls for far-reaching collective action to receive a wider hearing. At this opportune moment, proposals for a basic income guarantee (BIG) and reduced work week may well gain more traction.

It remains to be seen, of course, whether a BIG moment is close at hand, still years away, or never to be. As part of a three-pronged transformative strategy (displacement, radical reform, and disruption), socially conscious Christian leaders do well to raise their hands in favor of basic income and a thirty-hour work week *now* in an effort to change the conversation *before* automation kicks in even more forcefully, in part because societies plagued by high unemployment are less likely to sacrifice economic growth to environmental protection. Doing so will enable us to establish common ground and open up space for consideration of desirable, viable alternatives to the neoliberal status quo. The reframing strategy I am advocating aims to shift attention away from stale debates between liberals and conservatives toward everyone's felt experience of "future shock" (Toffler) and growing recognition that smart, disruptive technologies are leading to a jobless future for many. For Christian leaders, I wager the time has arrived to reframe the issues of poverty, rising inequality, and climate change as problems we must face together.

BIG Ideas

If the technological unemployment thesis holds water, then we can expect the question of how technology-driven productivity gains should be shared

---

35. Climate Leadership Council, "The Conservative Case for Carbon Dividends."

to become more acute as awareness of automation's impact grows. Questions of substantive freedom, social security, and well-being will become more salient as rising concern over technological unemployment creates an opening for a much-needed discussion around the terms and conditions for a just, humane transition to post-jobs society.

The political push for this transition must come from a broad social movement committed to "real freedom for all,"[36] i.e., a substantive freedom predicated on economic security (FDR's freedom from want) as a precondition for pursuit of the good life and human flourishing throughout the life cycle. While US political culture has become more polarized, still a rough consensus holds around the need to provide education, health care, food, housing, and other necessary goods to dependent groups (i.e., the young, old, and disabled), despite sharp differences in policy approaches. By contrast, public support for able-bodied adults remains more limited due in good part to the strong cultural bias toward formal employment as the preferred mode of responsible adult participation in society. However, in light of accelerating automation, this cultural bias—and the work ethic underwriting it—may become open to challenge.[37] Among bloggers and policy wonks we now find growing interest, both on the left and the right, in the BIG idea.

Sociologist James Hughes argues that a combination of rising technological unemployment and longer lifespans is creating a strategic opening for reconsideration of a universal, unconditional basic income. A political battle over entitlements is coming as less advantaged younger and middle-aged workers demand fairness from comparatively well-off seniors.[38] New technologies such as 3D printing and desktop manufacturing will eliminate much of the work done in between inventors and consumers at a time when the ratio of workers to retirees is shrinking. (This trend is global in scope; in China, e.g., currently there are five workers for every retiree, but the ratio will drop to 1.5:1 by 2040. It is no coincidence that China is at the forefront of robotics as it forecasts labor shortages resulting from its decades-long one-child policy.) Under these conditions, Hughes argues, "basic income is

---

36. I am adopting this slogan from Van Parijs, *Real Freedom for All*.

37. Kathi Weeks develops a critical analysis of how the work ethic legitimates industrial and postindustrial society's productivist bias in *Problem with Work*, chap. 1. Weeks usefully highlights the work ethic's continuing power as a discourse of disciplinary control (and increasingly, in a post-Fordist context, as a discourse of self-discipline for an expanding class of jobs now deemed "professional") as well as its multiple antinomies and vulnerabilities.

38. Hughes, "Strategic Opening."

the logical re-negotiation of the social contract to ensure that we don't spiral into widespread poverty and inequality."[39]

Consider Mark Walker and Andy Stern's basic income proposals as representative samples. Under Walker's scheme, every US citizen ages 18 to 64 (about 195 million total) receives $11,700 per year, with no means testing or work requirement. Walker's plan costs about $2 trillion a year, which is 11.5 percent of US GDP ($17.4T in 2014), and is paid for by a 14 percent value-added tax. This BIG proposal provides an income floor for every working-age adult, no strings attached, and in theory it yields net-income increases for those earning up to $80,000, which accounts for 90 percent of earners.[40]

Stern, a former president of the Service Employees International Union, suggests a similar figure of $12,000 per year for all US adult citizens. Under Stern's plan the disbursement amount is indexed to Gross Domestic Product (GDP), thus tying basic income levels to the nation's overall productivity. While Stern leaves open the question of how $2 to 2.5 trillion can be raised to fund a BIG, he does offer a menu of options in the hope that political compromise is possible. BIG funds might come from a combination of policies, including elimination of most federal welfare programs, doing away with market-distorting tax breaks, cancellation of proposed military spending for modernizing nuclear and other weapons systems as well as closure of numerous military bases overseas, and institution of financial transaction (Tobin) and wealth taxes (Picketty).[41]

As noted last chapter, I would add a carbon tax to the mix of possible funding sources. This option has the potential to raise a decent portion of the revenue required while also garnering bipartisan support, the continuing opposition from climate deniers in Congress and the Trump administration notwithstanding. Earlier I mentioned the proposal put forth recently by the Climate Leadership Council, a research and advocacy group organized by Republican leaders, for a carbon tax based on a fee-and-dividend model. The plan aims to achieve two goals: a market transformation to a clean, post-carbon economy and disposable income increases for most U.S. households. It bears repeating that the proposed carbon dividend is a basic income by another name, albeit a modest one. According to proponents, the moral case for a carbon dividend is based on just desert, as expressed in the polluter-pays-principle: those who continue high-carbon activities pay more in taxes (which increase over time), while those who adopt low- and

---

39. Ibid.
40. Walker, "BIG and Technological Unemployment."
41. Stern, *Raising the Floor*, chap. 8.

zero-carbon behaviors, investment strategies, and technologies are rewarded for their smart, responsible choices with a dividend (initially about $2,000 USD for a family of four).[42] Finally, the plan includes border carbon adjustments (rebates and tariffs) that encourage other countries trading with the US to adopt carbon taxes as well.[43]

Will a BIG undercut the work ethic or be wasted on alcohol, drugs, and frivolous spending? Pilot programs in several countries indicate that motivation to work does not weaken significantly with a basic income in place, and especially in developing nations it yields appreciable gains in income and well-being as recipients use the extra cash to make investments in the future. A 2009 experiment in London found that all thirteen "hard core" homeless men given a one-time payment of $4,500 took positive steps to improve their situation, moving off the street, taking classes, and entering drug treatment programs. None used the money to party on, and the project cost was dramatically lower than what welfare agencies had been spending, fruitlessly, year after year.[44] Recent research suggests that poverty-induced stress leads to poor decision-making.[45] Contrary to conservative claims that "free money" will lead to rampant laziness and drug abuse, the freedom from want afforded by a BIG enables the poor to make smarter, more responsible choices.

In OECD countries, the BIG idea finds support across the political spectrum for its streamlined approach to combating poverty and its respect for autonomy. Unlike means-tested programs, it "answers the Foucauldian critique about the welfare state being a way for the state to stigmatize and control marginalized populations," and also provides "a way of recognizing the value of decommodified caregiving and other cooperative, non-labor activities, by making sure there is space in the economy to both reward and

---

42. A Catholic concern to balance rights (distributive justice) with responsibilities (contributive justice) seems satisfied here. That said, I argue below that a more robust Catholic social-ethical argument for a larger BIG of $12,000 per year rests on a different moral logic and additional principles such as need, the universal destination of created goods, the social mortgage on private property, and the common good. A loving Creator intends the goods of the earth to satisfy the basic needs of all, yet access to productive property effectively is denied to large numbers of modern workers, who instead must sell their labor for wages. To date, modern Catholic social teaching has emphasized the duty of employers to pay workers a living wage as a way to satisfy basic needs and recognize labor's contribution to wealth creation, but the time has come to reassert an older, complementary argument for a just redistribution of wealth and economic opportunity through the mechanism of basic income.

43. Climate Leadership Council, "The Conservative Case for Carbon Dividends."

44. Bregman, *Utopia for Realists*, 25–27.

45. Mani et al., "Poverty Impedes Cognitive Function."

carry them out."⁴⁶ Labor activists highlight the potential of universal basic income to alter the balance of power in favor of workers, each of whom would now have a "permanent strike fund" (Stern) to resist the encroachments of aggressive managers. With bargaining power shifting toward labor, demands for better working conditions and more meaningful work no longer could be dismissed so easily. As Eric Olin Wright notes, "By increasing workers' capacity to refuse employment, basic income generates a much more egalitarian distribution of real freedom than ordinary capitalism, and this directly contributes to reducing inequalities in access to the means to live a flourishing life."⁴⁷ On this view, a BIG sets the stage for a more democratic (as opposed to corporatist) co-determination of employment's terms and conditions across the economy.

Philosophically, a basic income can be defended on both communitarian and libertarian grounds.⁴⁸ Communitarians supportive of a BIG emphasize the desirability of a society in which all citizens can participate with dignity in building up the common good. The dignified social participation made possible by a basic income enhances the quality of community life and helps to create a more stable, high-trust culture. With freedom from want established and more time available, individuals are empowered to contribute their time and talent to family and community. For their part, libertarians look favorably on a BIG that eliminates welfare-state dependency and expands personal freedom to choose forms of participation in the economy that make sense for one's life situation. Since a basic income by design only meets essential needs at a modest level, it need not undermine work incentives or generate an unmanageable free-rider problem. As a practical matter, most adults would supplement their basic income with paid work, only now they would be in a position to make thoughtful choices about how much and how hard they work rather than feel compelled by economic necessity. The "bottom line": freedom from want *empowers* people to pursue good lives.

From a Catholic perspective attuned to the plight of the poor yet wary of socialist solutions, a basic income offers a practical way in a liberal polity to balance respect for individual rights and freedoms with the demands of the common good. In developed countries where an excluded minority caught in a poverty trap faces a resistant coalition of the better-off and out of frustration turns to deviance and violence, a basic income offers society

46. Konczal, "Thinking Utopian."
47. Wright, *Envisioning Real Utopias*, 218.
48. See the various essays collected in Widerquist et al., *Basic Income*. In the US, Christian social-ethical support for the BIG idea first appears in King, *Where Do We Go from Here*; see also Wogaman, *Guaranteed Annual Income*.

a democratic alternative to further mistrust, instability, and polarization. Bill Jordan argues that a basic income would eliminate the poverty trap associated with means-tested welfare and "give the excluded minority access to the market system of fairness by reward for effort, and hence to savings, property and other private goods. But also, by giving everyone a universal share of resources on the grounds of membership (citizenship), it would be a mechanism for including all in the common good."[49] Along with universal health care and education, a well-designed BIG offers a viable option for realizing the Catholic moral vision of a good society that protects the vulnerable and empowers the poor to participate with dignity in social life, develop as persons, and contribute to the common good.

Christians believe a loving Creator intends the goods of the earth to satisfy the basic needs of all. An important yet neglected doctrine, the universal destination of created goods establishes the value and legitimacy of the commons (or public goods) and also places private property under a social mortgage. In other words, both personal and privately-held productive property, while legitimate and useful for the healthy functioning of a market economy, must be used responsibly for the benefit not only of its owners but also in ways that contribute to the well-being of the community and its less-advantaged members. For example, in situations such as Brazil where the basic needs (e.g., food and water) of many go unmet, agricultural land may be appropriated justly from landowners with large, unproductive holdings. More broadly, where historic injustices (i.e., aristocratic enclosures, colonial expropriations, and unregulated monopolies) lead to property regimes and distorted ownership patterns that concentrate wealth (or capital) and effectively deny large numbers of modern workers access to productive property, some mechanism for redistribution is both necessary and just, since a wider and more equitable distribution of wealth and economic opportunity is an important, structural component of the common good in any modern, democratic society. By contrast, highly unequal societies are prone to social unrest, political instability, authoritarian rule, and high levels of repressive and revolutionary violence. In a slogan: "No economic justice, No social peace!"

To date, modern Catholic social teaching since Pope Leo XIII has dealt with the imbalance of power between capital and labor by emphasizing the duty of employers to pay workers a living wage as a way of satisfying the basic needs of worker households, recognizing labor's contribution to wealth creation, and offering frugal workers the opportunity to save and invest in wealth-generating resources (e.g., a parcel of land, a house, stock holdings).

49. Jordan, "Basic Income."

In the context of an industrial society characterized by Fordist production, the church also upheld the right of workers to organize unions and collectively bargain for a just share of socially-produced wealth. However, from the early 1970s onward, the context for evaluating questions of economic justice and power has been changing: owners and managers of capital have responded to rising costs and declining profits with a neoliberal strategy that flexibilizes and atomizes the labor process, erodes welfare-state protections, reduces or eliminates company benefits, automates and/or outsources jobs aggressively, and thus shifts the *risks* associated with unemployment, poverty, and debt onto workers. As noted earlier, both the (self-identifying) middle class and working poor have registered these changes in high levels of stress and economic insecurity. In recent decades the "precariat" (Standing) has grown, while safety-net programs designed for a twentieth-century economy have not been modernized. Given this new reality, the time has come for the church to rethink its emphasis on a living wage and union formation as the primary policy tools for empowering workers. Today, the wider distribution of wealth and economic opportunity called for in Catholic social teaching can be achieved more effectively through additional policies, namely a basic income and thirty-hour work week. In sum, the BIG idea provides a way to implement the church's longstanding teaching on the universal destination of created goods, the social mortgage on property, and the demands of the common good. It also strikes the right balance between the twin principles of solidarity and subsidiarity, since it builds a solid floor of social support for all while at the same time enabling (and expecting) each recipient to build their own future (rather than being "nudged" by bureaucrats into "correct" behaviors).

By transitioning over time to a diverse, hybrid economy, American society can achieve higher levels of economic and social security across all ages without sacrificing the vitality and dynamism characteristic of (reformed) markets. Supported materially by a twenty-first-century smart infrastructure and socially by a basic income and other components of a well-twined safety net, this new economy will distribute productive capacities and economic opportunities more broadly across society. Crucially, a BIG need not foster indolence, dependence, and mindless consumption. With the rapid advance of affordable 3D printers and other DIY technologies, we have noted, millions of aspiring artisans and entrepreneurs will possess or enjoy cheap access to sophisticated means of production otherwise out of reach. Etsy, TechShop, Columbus Idea Factory, and other maker-spaces dedicated to growing the artisanal economy present working models. By establishing real freedom for all, a BIG would allow the next generation of innovators to risk new ventures without fear of economic freefall.

The rise of distributed manufacturing and self-provisioning makes the transition to post-jobs society a concrete-utopian option rather than fantasy.[50] In *Mass Flourishing*, Edmund Phelps describes how a dynamic US economy during the nineteenth century fostered innovation on a wide scale: "Even people with . . . modest talent were given the experience of using their minds: to seize an opportunity, to solve a problem, and think of a new way or a new thing."[51] Combined with a thirty-hour work week, a BIG would help to create more favorable conditions for a revival of "indigenous innovation" (Phelps) through the pursuit of more engaging, meaningful work. Such pursuits move from day-dream to real possibility with the expansion of a civil market economy and release of millions of workers from the soft tyranny of full-time wage labor.

Liberated from fear of want, it becomes easier to contemplate the simple and quite powerful question asked by social innovator Frithjof Bergmann: What do you *really* want to do? With successful demonstration projects in Michigan (Flint, Detroit) and elsewhere, Bergmann and the "new work, new culture" movement show that asking people what they really want to do, and then supporting their deepest aspirations with coaching and training, is a powerful formula.[52] More broadly, as the viability of new-

---

50. For case studies of the new economy in the making, see Schor and Thompson, *Sustainable Lifestyles*.

51. Phelps, *Mass Flourishing*, 15. Phelps pleads for a renewal of modern, take-off economies driven more by "indigenous innovation" than science, pitting their dynamism and widespread prosperity against "traditional" economies more concerned with stability and community cohesion and their twentieth-century counterparts, the "hobbled" EU social democracies. He argues that a modern economy, provided it is governed justly by Rawlsian policies and governing institutions that work to the benefit of the least advantaged, offers the best hope for reviving the "heroic spirit" of indigenous innovation, which he associates with classic conceptions of the good life centered around engagement, challenge, and the extension of human capabilities. While I second the call for a revival of indigenous innovation and engaging work, the new-distributist program I am advocating differs considerably from Phelps's proposal in taking technological risks and ecological limits more seriously. The technological *and* social innovation we need must be directed to sustainable abundance for all, not affluence for an overly privileged few.

52. Bergmann and Staehelin, *Starting With New Work*. Along similar lines, Holt points to the limits of a sustainable economy movement strategy focused on communications around global issues, bold policy proposals, and examples of social enterprises that cater to an upscale, culturally liberal consumer market concentrated in large urban areas. A progressive, post-scarcity, cosmopolitan vision does not resonate culturally with tens of millions of US workers who do not hold college degrees, are unemployed, underemployed, or stuck in insecure and low-paying jobs, and presently do not have the skills and/or working capital required to start their own businesses. Rather than communications, activists need to prioritize creation of innovative demonstration projects in local communities that offer work opportunities and affordable services for

work projects and practices of self-provisioning becomes known, public support for policies supportive of an expanding civil market economy (e.g., public funding for community-based new-work centers) will grow, creating momentum for a continuing transition to a diverse, hybrid economy and post-jobs society oriented to sustainable abundance for all.

A new distributist policy agenda also includes requiring companies to reduce work hours without pay reductions (or with a modest reduction where necessary), which amounts to a de facto redistribution of productivity gains arising from innovation and automation. Reducing the work week to thirty hours can and should be a major objective of a renewed labor movement. Progressive businesses in the EU and elsewhere have experimented successfully with shorter and more flexible work hours; employee morale and productivity typically are higher in these workplaces.

All this lies within the realm of the adjacent possible, even if the present constellation of political forces within the US and elsewhere renders this project implausible to many at first blush. Granted, the "unthinkable" at present is *post-jobs society*, a really big idea pointing toward a future that involves much more than achieving the utilitarian goal of steadily increasing productivity and higher living standards within the current system. Can we envision and work to organize a broad social movement pushing for a just, humane transition to a diverse, hybrid economy and socially inclusive post-jobs society in which full-time wage labor no longer sets the parameters for life's meaning and value?

Such a movement—at once political and cultural in character—begins in earnest with the refusal to work without end. With Antonio Negri and other autonomist Marxists, Kathi Weeks defends the refusal to work—i.e., the refusal to make holding a full-time job the central focus of individual identity and social life—as a labor strategy that advances a critique of the wage system while also opening a space of freedom for new identities, alternative communities, and social innovation beyond monopoly capitalism. In her view, "the refusal to work serves not as a goal, but as a path—a path of separation that creates the conditions for the construction of subjects whose needs and desires are no longer as consistent with the social mechanisms within which they are supposed to be mediated and contained . . . The defection enacted through the refusal to work is not predicated upon what we lack or cannot do, it is not the path of those with nothing to lose but their chains; it is predicated instead on our 'latent wealth, on an abundance of

---

this main street audience; see Holt, "Why the Sustainable Economy Movement Hasn't Scaled." Absent these practical alternatives, large segments of this demographic will remain politically disengaged or be susceptible to a right wing politics of fear and resentment, as the recent Trump victory attests.

possibilities."[53] Realizing these abundant possibilities will require enactment of transformative strategies (displacement, radical reform, and disruption)[54] as well as cultural renegotiation of the work ethic. We can anticipate that openness to and readiness for systemic change will grow as more workers are impacted by accelerating automation and the (self-identifying) middle class comes to terms with the root causes of poverty, rising inequality, and structural unemployment. In this scenario, a new distributist program of radical reform—anchored by a basic income and shorter work week—could be viewed as a practical requirement underwritten ethically by the demands of distributive justice and motivated by a general interest in maintaining social peace within an increasingly divided society. And for the idea and ideal of post-jobs society to take hold, cultural attitudes toward work and leisure need to change as well. Before we explore how that might happen, let us consider more closely what a future post-jobs society might look like.

A Really BIG Idea: Post-Jobs Society

A basic income, reduced work week, and other policies such as universal health care and education would enable the US to transition to a diverse, hybrid economy and post-jobs society, i.e., an economy and society not predicated on the goal of full employment and norm of a forty-hour work week. In post-jobs society a person's identity, social status, self-esteem, and well-being will be de-linked from wage labor. People will enjoy the substantive freedom and security to fashion life-plans characterized by a more flexible mix of paid work, self-provisioning, family/household duties, personal interests, community service, and political activity. With less time spent on the job, more time can be devoted to engagement with focal concerns that contribute to human flourishing. While larger financial rewards will continue to go to those who choose to take on more demanding professional and technical work or pursue high-performing roles requiring a greater career commitment, in post-jobs society such individuals and groups will not be overly privileged. A post-jobs society will develop ways to recognize and reward a wider range of contributions to society without undercutting

---

53. Weeks, *Problem with Work*, 100.

54. Much more is involved than strikes by organized labor, though the strike will remain an important disruptive tactic available to workers. Besides disruption, there is the gradual displacement of neoliberal ideas and institutions from their hegemonic position through the buildup of alternative institutions and practices. Here the "refusal to work" (i.e., accept wage labor on nearly any terms out of desperation) has many practical meanings.

standards of excellence and incentives to high achievement in various fields of endeavor.

Along these lines, Ulrich Beck proposes a post-jobs model: the "multi-activity" society based on political freedom and civil labor supported by a state-funded social wage (i.e., some form of a BIG): "Alongside paid work, [civil labor] constitutes an alternative source of activity and identity which not only gives people satisfaction, but also creates cohesion in individualized society by breathing life into everyday democracy . . . Civil labor is voluntary, self-organized labor, where what should be done, and how it should be done, are in the hands of those who actually do it."[55] Beck emphasizes the need for social entrepreneurs who spearhead civil-society initiatives that address community needs unmet by government. His vision of a multi-activity society allows us to counter worries about over-entitlement, indolence, and government bureaucracy. Cultural suspicion of the "vice of idleness" can be allayed, for a post-jobs future will be one in which the structures and policies of society encourage and materially support formal employment, civil labor, and self-provisioning while also providing more time for family, community, rest, recreation, culture, and celebration. The latter activities go unremarked by Beck, but I argue below that a desirable post-jobs society and culture calls for re-balancing work with enjoyment of fruitful leisure.

Post-jobs society means less time on the job, coupled with an economic security that enables us to discover and pursue what we really want to do; economic security opens up opportunities to experiment with roles and activities beyond wage labor. From a theological-ethical perspective, a vision of post-jobs society aims at the practical realization of substantive freedom (as opposed to the empty, formal freedom and "consumer sovereignty" of liberal society), which is a precondition for pursuing good lives that convivially co-create greater value and beauty.

Without question, resistance to the "radical" idea of post-jobs society will be difficult to overcome, given the present state of US political culture. However, as the authors of "The Post-Work Manifesto" (1998) state, "To imagine a different way is always a risk. But in the world where work is destroyed by global capital and computer-aided technologies, not to imagine an alternative is to take the greater risk."[56] As we articulate a vision of sustainable abundance for all and advocate for policies that enable a just, humane transition to a socially inclusive post-jobs society, we also need to rethink and recover Catholic social teaching on work and leisure.

55. Beck, *Brave New World of Work*, 127.
56. Aronowitz and Cutler, *Post-Work*, 70.

The Priority of the Cultural

How do we spend our time? And what degree of control do we have over our hours and days? These are profoundly ethical questions. As discussed in chapter 5, our complicity with the device paradigm makes moving beyond live-to-work habits and commodified leisure a difficult challenge, especially given the cultural hold of the work ethic and sophistication of the culture industry. For Catholics, the challenge of putting work—and the work ethic—in its place and reimagining leisure will require a revision and retrieval of church teaching.

As with John Paul II in *Laborem exercens*, Benedict XVI and Francis ascribe a twofold meaning to work. First, human toil is always in some respects a "curse" stemming from the Fall, an onerous and unavoidable task that invariably strains both body and psyche. Second, work provides opportunities for self-development and fraternity; this is the personalist dimension of work that grounds, e.g., John Paul's notion of the priority of labor over capital and critique of alienating, exploitative forms of work. A certain tension exists between these two meanings, a realm of necessity at odds with a realm of freedom. Yet there is more at work here than meets the eye, for modern Catholic social teaching on work's twofold meaning is not derived solely from biblical or personalist-philosophical sources. Rather, the meaning of work is disclosed within the horizon of industrial capitalist society, where a great deal of work takes the form of alienated, exploited, and slave labor,[57] and where blue-, pink-, and white-collar workers in a production-obsessed culture are conditioned to find their identity, status, and self-worth primarily through wage labor.

Recent papal social teaching has not addressed the issue of technological unemployment adequately, in my view. In *Caritas in veritate*, Benedict calls on governments to take measures necessary to ensure full employment under dignified conditions. While Francis takes note of automation's negative impacts in *Laudato*, he says much the same thing and does not contemplate the possibility of a transition to post-jobs society. However, in the not-so-distant future Catholic social teaching will need to rethink the relationship between work and leisure within a new context, namely an emerging society in which labor of many kinds is outsourced to smart machines. To be sure, the personalist meaning of work will retain its force as a powerful moral critique of neoliberalism and technology-centered automation.

---

57. On modern slave labor and its devastating environmental impact in Brazil, eastern Congo, Bangladesh, and many other locales in the global South, see Bales, *Blood and Earth*.

At the same time, we must challenge the overvaluation of (paid) work itself, and not simply the alienation and exploitation of workers.

Here a retrieval of the church's social wisdom on the right use of leisure is in order. As we saw in chapter 3, *Gaudium et spes* affirms that, in addition to a right to dignified labor, workers should also "enjoy sufficient rest and leisure to cultivate their familial, cultural, social and religious life. They should also have the opportunity freely to develop the energies and potentialities which perhaps they cannot bring to much fruition in their professional work" (GS #67). A politics of engagement and time places a priority on pursuing good lives beyond the banalities of so much mindless wage labor; it places a premium on maximizing opportunities for channeling "energies and potentialities" toward focal practices that constitute human flourishing. This is really what the people-over-profits economy envisioned by Catholic social teaching promises.

That said, the creation of more paid work will be necessary (in the capitalist and civil market sectors primarily, but also through government programs, e.g., a modernized WPA managed online), since even with a BIG in place and self-provisioning widely practiced, people still will need additional income to pay for necessities, create savings, and enjoy leisure. In addition, the demand for more meaningful work within the formal economy must be put on the table if we are to address the quiet crisis of widespread employee disengagement. Just as vital, though, is a cultural shift supportive of a broad social movement to put paid work—and the work ethic—in its proper relation to life.

Post-jobs society promises to open up new and expanded spaces not only for practices of democratic self-governance, civil labor, and social entrepreneurship, but also for various types of fruitful leisure, which together become an important counterpoint to modern work compulsions, on the one hand, and the sad reduction of leisure to the sensate and superficial, on the other. The experience of *deep play*, for example, refers to our intense engagement with life through the arts, games, and (pure) sciences. As Diane Ackerman puts it, "Deep play is the ecstatic form of play. In its thrall, all the play elements are visible, but they're taken to intense and transcendent heights."[58] Our avocations matter, and in a post-jobs society with basic income security we can pursue these passions more fully without fear or guilt.

Fruitful leisure also includes participation in "communities of celebration" (Borgmann) as well as cultivation of various spiritual practices. At its best, the experience of liturgy takes worshipers in a spirit of play beyond themselves—i.e., beyond their individual preoccupations—into the mystery

---

58. Ackerman, *Deep Play*, 12.

and joy of divine presence. In *The Spirit of the Liturgy*, Romano Guardini recommends entering into prayer and worship in a spirit of playfulness:

> The soul must learn to abandon, at least in prayer, the restlessness of purposeful activity; it must learn to waste time for the sake of God, and to be prepared for the sacred game with saying and thoughts and gestures, without always immediately asking "why?" and "wherefore?" It must learn not to be continually yearning to do something, to attack something, to accomplish something useful, but to play the divinely ordered game of liturgy in liberty and beauty and holy joy before God.[59]

Leisure becomes spiritually fruitful whenever our capacity for contemplation, awe, and wonder awakens and grows; it occurs whenever we shift to receptive, non-instrumental modes of consciousness that allow us to encounter divine mystery and connect at the core of our being with the sacred within and all around us.

Experiences of fruitful leisure are sources of the self's sense of psychic integrity, of wholeness and peace, of profound gratitude and joy. In these moments we sense our envelopment within an unfolding plenitude so great as to defy description. Beyond our passing enthusiasms and optimistic moods, what energizes and inspires us most fully, what provides us with a sense of direction in the movement of life, are these encounters with superabundant grace. Through varied forms of fruitful leisure we discover the ultimate ground for every step we take toward creating a world of sustainable abundance for all. We come to understand why the political struggle for a just, humane transition to post-jobs society is worth the effort.

As Catholic philosopher Josef Pieper has taught us, leisure is the basis of culture. Fruitful leisure is integral to the flowering of democratic culture, and it is a vital element of any credible and convincing conception of the good life. (Its opposite, a decadent leisure that renders us passive consumers of corporate products, must be resisted.) Fruitful leisure is a necessary complement to Beck's vision of political freedom and reinvention of everyday democracy through civil labor and social entrepreneurship. The immense good of fruitful leisure is grounded in anthropology and theology: through deep play we express ourselves as *homo ludens*, and activities of fruitful leisure make good on our calling to be celebrants of God's creation (*homo adorans*). Chapter 10 explores further the significance of deep play and other forms of fruitful leisure as antidotes to the alienations of technological society.

---

59. Guardini, *Essential Guardini*, 152.

However, we can expect real difficulties in the transition to a post-jobs society in which wage labor no longer defines our existence. To state the challenge simply: With more time on our hands, will we know what to do with ourselves? At face value the question seems ludicrous, but when we consider the cultural resonance of the work ethic, on the one hand, and how extensively commodified leisure has become, on the other, the force of this query quickly returns.

The advent of automation, Hannah Arendt argued in *The Human Condition* (1958), had put the age-old dream of freedom from labor's toil and trouble within reach. Yet the possibility of a post-jobs existence, she noted, was self-defeating and threatening to the work-obsessed culture of modernity: "The modern age has carried with it a theoretical glorification of labor and has resulted in a factual transformation of the whole of society into a laboring society . . . It is a society of laborers which is about to be liberated from the fetters of labor, and this society does no longer know of those other higher and more meaningful activities for the sake of which this freedom would deserve to be won."[60] Beyond the labor required for reproduction and material comfort, moderns know only commodified leisure. The distinctive activity of politics, Arendt lamented, is now reduced to another job category.

While Arendt's prediction of a post-jobs society was premature, the "leisure question" she posed is worth asking anew in light of accelerating automation: Are industrial moderns capable of taking up "those other higher and more meaningful activities" that make freedom from wage labor a genuine advance? Her contemporary Pius XII believed so. To recall the discussion from chapter 3, in 1957 the "pope of technology" delivered an address in which he acknowledged automation's potential to free up time for workers, who then would face the challenge of using leisure wisely. A right use of leisure required development of the self's intellectual and spiritual life, and the pope hoped that with more leisure time "the most profound needs of the soul will find their satisfaction." However, if men and women "yield to the temptation of a life of greater ease filled with more and more sensual pleasures, [they] will gain nothing from it except another kind of slavery and a certain moral decadence." Pius understood precisely, as did Arendt, the issue at stake in a society shaped by the pattern of technology.

---

60. Arendt, *Human Condition*, 4–5. Weeks echoes the point: "The glorification of work as a prototypically human endeavor, as the key both to social belonging and individual achievement, constitutes the fundamental ideological foundation of contemporary capitalism: it was built on the basis of this ethic, which continues to serve the system's interests and rationalize its outcomes. The contemporary force of this code, with its essentialism and moralism of work, should not be underestimated" (*Problem with Work*, 109).

Where technology promised both freedom from hunger, disease, and toil and freedom for self-development, the irony of modern technology is the creation of mindless labor and commodified consumption—the device paradigm.

Arendt's prescient observations regarding the eclipse of participatory democracy as well as modernity's attenuated capacities for fruitful leisure should give us pause. The modern overvaluation of (paid) work and lapse into passive, enervating leisure has cost moderns dearly in three ways: one by devaluing Arendtian action in the public sphere, two by neglecting traditions, skills, and habits of mind and heart associated with focal practices, and three by denying people the psychic health, ethical sensitivity, and spiritual awareness derived from fruitful leisure. Facing up to this predicament and changing our ways is what I mean by the *priority of the cultural*. Given the cultural hegemony of the work ethic and cult of entertainment, it will take nothing less than a trans-valuation of values over time to realize the full potential of a post-jobs society.

Along with Arendt, Borgmann, and others, Weeks has taken the measure of what we are up against: "Although there is no scarcity of possible reforms that could help us better to cope with the problems of unemployment, underemployment, precariousness and overwork in the contemporary economy—a shorter legal working day and a guaranteed basic income are two—the gospel of work and its central teaching, the work ethic, have so colonized our lives that it is difficult to conceive a life not centered on and subordinated to work."[61] Cultivating a genuinely good life beyond wage labor (and for many, it amounts to wage slavery), that's the priority of the cultural in a nutshell. It involves, among other things, an entirely new sense of work-life balance, development of self-provisioning skills, new gender relations, and bold experimentation with alternative, non-commodified models of community (e.g., co-housing) and the good life (e.g., voluntary simplicity). Family life will be an important site of struggle and emancipation within a post-jobs society in the making. The priority of the cultural begins at home, as it were. We can hope for a fulfilling post-jobs life only if married partners are able to rebalance roles and power within the household.

Twin Priorities

To sum up, the other inconvenient truth of technological unemployment presents Catholics and other persons of faith with an opportunity to change the conversation and reframe debates over poverty, inequality,

---

61. Weeks, "Imagining Non-Work."

automation, and climate change. As automation accelerates, Christian leaders do well to join with others in advocating for new distributist policies such as basic income and reduced working hours. A Catholic social vision of sustainable abundance for all calls for engagement in the political struggle for a just, humane transition to a socially inclusive post-jobs society. Moreover, this "priority of the political"[62] must be matched by commitment to the priority of the cultural. The latter involves learning how to step off the treadmill of production and consumption so that we can pursue meaningful work, strengthen family life, participate in communities of focal practice, and enjoy fruitful leisure more fully. Good lives are at stake.

The refusal to work without end marks the beginning of the broad social movement and long-term cultural shift we need. Persons of faith who grasp the significance of attending continually to divine reality, and thus not allowing capitalist powers and principalities to dominate their psychic and social reality, can and should lead the exodus from drone-work days and Saba nights. In the final chapters, I explore further how commitment to these twin priorities may open pathways toward a world of sustainable abundance for all.

---

62. In the ongoing debate over why some countries prosper and others stagnate or decline, Acemoglu and Robinson argue the key factor is neither geography nor culture. Rather, collective action and political struggle aimed at reforming institutions and establishing a functional democracy (pluralism, rule of law) is the decisive factor. When political institutions become more inclusive (e.g., the Glorious Revolution in seventeenth-century England), they generally support the emergence of more inclusive economic institutions, which in turn enable more groups to gain an "ownership stake" in society and support the expansion of democratic rule and political participation in a virtuous circle of sustained, broad-based prosperity; see Acemoglu and Robinson, *Why Nations Fail*, chap. 11. Cf. Hardt and Negri, *Multitude*.

9

Technology, Up-Down Politics, and the Creative Middle

The green argument that we only have one world and that it is each generation's duty to hand it on to the next as a sustainable social and natural habitat is no longer the preserve of utopians and environmental zealots; it is the new common sense before increasingly dangerous environmental risks.

—ANTHONY GIDDENS AND WILL HUTTON

People often go through three stages in considering the impact of future technology: awe and wonderment at its potential to overcome age-old problems; then a sense of dread at a new set of grave dangers that accompany these novel technologies; followed finally by the realization that the only viable and responsible path is to set a careful course that can realize the benefits while managing the dangers.

—RAY KURZWEIL

Any technical solution which science claims to offer will be powerless to solve the serious problems of our world if humanity loses its compass, if we lose sight of the great motivations which make it possible for us to live in harmony, to make sacrifices and to treat others well.

—POPE FRANCIS

What common sense calls for in a world of increasingly dangerous technological and environmental risks is not always self-evident, but it is clear that political battles are intensifying over how various threats to human and ecological health are defined and distributed in world risk society. As the "manufactured uncertainties" (Giddens) generated by industrial modernity proliferate, people and communities sense their vulnerability and lose confidence in the ability of scientific and legal authorities to ensure safety. Amidst growing demands for transparency and accountability, a self-endangering progress creates an ongoing legitimation crisis and thus opens up opportunities for social movements and civil-society actors to influence public policy.

In chapter 2, I noted how policy-makers and ordinary citizens alike have become more aware of a forced belonging to risk communities whose fate remains uncertain. We don't know where or when the next black swan event—the very improbable, high-impact events that shape history—will occur. Denial of black swan blindness aggravates our predicament further. That said, we do know that many risks to human and environmental health stem from the unintended, second- and third-order consequences of widespread technology adoption. A ubiquitous technology can trigger black swan events, such as the thinning of the ozone layer once CFCs went into general use after World War II. Moreover, if nanotech, synthetic biology, robotics/AI, and other emerging technologies prove capable of developing at an exponential rate, as Ray Kurzweil and others forecast, then the systemic and existential risks attending them will grow dramatically.[1] In short, not only are we bad at anticipating black swans, we also seem to have arrived at a moment in history when—for good or ill—technology-driven black swans may appear much faster and more often.

Powerful new technologies are part of a larger story. From the mid-eighteenth century onward modernizing societies have been caught up in a self-propelling process of technological, social, and cultural acceleration, or what Hartmut Rosa refers to as the "circle of acceleration." According to Rosa, global society has moved quickly since 1990 into a phase of hyperacceleration characterized by a paradoxical state of "frenetic standstill." Under conditions of hyperacceleration, politics tends toward a muddling through from one emergency to the next. Hyperacceleration thus undermines the prospects for deliberative democracy to the extent that pressure continuously mounts for decisions to be made quickly and with only short-term consequences in view. As risk conflicts proliferate and technology development accelerates, a new up-down contest between techno-progressives

---

1. See Bostrom and Cirkovic, *Global Catastrophic Risks*.

and bio-conservatives is emerging. The former seek to advance a proactionary agenda of exponential technology development and human enhancement, while the latter advocate a strong precautionary approach to new technologies.

This chapter explores two questions broached earlier: How do we assess technologies when there is scientific uncertainty regarding potential impacts that could prove catastrophic and irreversible? How might policymakers, civil-society groups (including faith-based groups), and concerned citizens engage with and respond to techno-progressive and bio-conservative perspectives in public debates over technology's trajectories? After presenting a case study on energy options in south Florida that highlights some of the issues attending world risk society, I turn to the questions at hand. Drawing on the writings of Kevin Kelly, I first propose an *evolutionary ethics of responsible care* that is available in principle to all persons, consistent with an eco-theological vision of sustainable abundance for all, and capable of addressing realities on the ground. I then explore how a creative-middle approach informed by Bernard Lonergan's heuristic account of the dialectics of history can help us avoid the Scylla of recklessness and Charybdis of relinquishment in technology development. By way of conclusion, I situate the proposed evolutionary ethics of responsible care and creative-middle approach within the context of a radical reform of technology.

The Days before Thanksgiving at Turkey Point

Operated by Florida Power & Light, Turkey Point Nuclear Generating Station sits twenty-five miles south of Miami along the shore of Biscayne Bay. Turkey Point's five power units, which supply nearly all of south Florida's electricity, include two aging nuclear reactors (Units 3 and 4) that in 2002 were granted operating license extensions, from forty to sixty years, by the Nuclear Regulatory Commission (NRC). Projecting increased demand, Florida Power & Light submitted a request to the NRC in 2007 to build two new nuclear reactors (Units 6 and 7) at Turkey Point at a projected cost of $26 billion, with ratepayers footing the bill ($281 million collected from the former through early 2017).

Opposition to Florida Power & Light's plan arose early from local politicians and environmental groups and has grown in recent years. Critics cite several concerns: water quality and use issues affecting local groundwater, Biscayne National Park, and the Everglades; increased risks from rising sea levels and stronger hurricanes; a possibly flawed nuclear reactor design as well as insufficient evacuation zones; and a culture of silence among workers at Turkey Point.² These issues—and the politics surrounding them—light up the challenges confronting us in a risky, runaway world.

Keeping nuclear reactors cool enough to operate efficiently and safely requires huge volumes of water. Turkey Point's closed-loop system is unique among US nuclear plants. Rather than using towers for cooling, it circulates discharged water through a 168-mile-long series of open-air canals (from the air, it looks like a giant radiator) and then reuses it for cooling within the plant. In recent years the heavy, hyper-saline water has helped push an underground saltwater plume further inland toward drinking water supplies.³ During this time, the canal system also has suffered from large algae blooms. Along with periods of low rainfall, the heat-trapping algae have caused water temperatures to rise to dangerous levels, threatening a shutdown of plant operations. In 2014, canal water temperatures exceeded the 100-degree limit set by law. With the NRC's blessing, Florida Power & Light was allowed to operate the plant's cooling system temporarily beyond the 100-degree safety limit, up to 104 degrees if necessary.

Beyond pleading for emergency waivers to nuclear safety laws, Florida Power & Light's response to these issues has been to apply a chemical treatment to reduce the algae and to access, on an emergency basis, water from the Biscayne Aquifer and L31-E canal (part of the vast Everglades water-control system) to cool down and reduce the salinity of its canal system water. If algae blooms and high saline levels persist, then the utility's need to tap these other sources for cooling and dilution will compete directly with the drinking water needs of residents in south Florida as well as ongoing restoration efforts in Everglades National Park.

Florida Power & Light says its plan for cooling the new reactors, using towers this time, will improve water quality. The new system will use reclaimed wastewater from a nearby county sewer plant, with saltwater from nearby Biscayne Bay used as a backup. The company also plans to inject wastewater from new nuclear operations into salt-water caverns deep underground, a disposal option that raises concerns about eventual seepage of contaminated water into local freshwater aquifers. In response, local mayors and other critics point to the investor-owned utility's poor track record in managing the forty-year-old canal system. A persistent pattern of evasion, denial, inaction, and resistance to monitoring and research arouses suspicion that short-term profits are being put before long-term commitments to human and environmental health. They question how safely sequestered the

---

2. Other substantive issues include the proposed project's high cost, which ratepayers will absorb rather than investors, and FLP's plan to construct an overhead high-voltage power distribution system instead of safer, underground transmission lines.

3. Staletovich, "State eases oversight."

new wastewater will be, even if it's deposited way below freshwater sources located closer to the surface.[4]

Meanwhile, managers of Biscayne National Park, a marine sanctuary located about ten miles due east of Turkey Point—worry that Florida Power & Light's operations could impact already stressed coral reefs and aquatic life. As ocean water temperatures continue to climb due to climate change, any changes to bay waters from Turkey Point discharges—e.g., contaminated wastewater seeping eastward from the unlined cooling canals into the bay—introduce an additional risk factor. A report released in March 2016 found that tritium, a radioactive isotope of hydrogen, has leaked from the canals into local groundwater and the bay. While tritium levels are low, biologists worry that canal water contamination may elevate ammonia and phosphorus levels, which in turn could harm the bay's marine life.

Despite these and other concerns, the NRC's final environmental impact statement (EIS) concluded in October 2016 that no adverse impacts of significance should prevent issuing licenses for the two new reactors at Turkey Point. For twelve resource categories (natural and human/cultural, including radiological health and environmental justice), the assessment ranks the expected impacts as small, moderate, or large. The key distinction is between moderate impacts that "alter noticeably, but do not destabilize, important attributes of the resource" and large impacts that "are clearly noticeable and are sufficient to destabilize important attributes of the resource." The EIS also weighs "cumulative impacts resulting from the proposed action when the effects are added to, or interact with, other past, present, and reasonably foreseeable future effects on the same resources." According to the NRC review team, construction and operation of the additional nuclear units will result in mostly small and only some moderate impacts to natural and human/cultural resources within the area. No large impacts, single or cumulative, are anticipated by the NRC (though mention is made in passing of a potentially large impact on terrestrial ecology resources over the long run from overall development in south Florida).

The surrounding Miami-Dade County population has grown significantly since Turkey Point began operating its nuclear units in 1972. NRC emergency planning designates two danger zones ringing nuclear power plants, one a plume exposure pathway zone with a ten-mile radius that deals with exposure to and inhalation of airborne radioactive contamination, the other an ingestion pathway zone, fifty miles in radius, which is concerned with radioactively contaminated foods and liquids. Over 160,000 people now live within ten miles of Turkey Point, and nearly 3.5 million reside

---

4. Staletovich, "Mayors make case."

within fifty miles. Critics charge that too many people inhabit these danger zones, and that feasible evacuation options fall far short of what is needed for moving a large urban population to safe areas in a timely manner if a major incident occurs.

A related concern has to do with the nuclear plant's bayside locale and vulnerability in the face of rising sea levels, more severe weather events, and stronger storm surges. Some 2.4 million area residents inhabit terrain just 4.4 feet above the high-tide level, and by 2050 the coastal region is likely to see storm surges on top of sea level rise reaching well above that level, according to recent climate-science forecasts. While affluent enclaves such as Miami Beach pass $500 million bonds to erect safeguards, Shorecrest and other communities of color in Miami-Dade County already face more frequent flooding from rainstorms.

In 2010, four of the region's counties formed a compact for long-term collaboration on climate-change adaptation. Their 2012 climate action plan is comprehensive in scope and includes 110 action items. Notably, the plan calls for a regional and state-level commitment to greater energy efficiency and deployment of alternative/renewable energy technologies supported by a renewable energy portfolio standard, new financing options, and changes in building codes, land-use rules, and architectural designs that enable locally generated renewable energy to flow efficiently into the power grid.[5] For its part, Florida Power & Light has reminded detractors of how the containment buildings at Turkey Point withstood a direct hit from Hurricane Andrew (Cat-5) in 1992, and says it plans to build Units 6 and 7 on reinforced platforms at twenty-six feet above sea level, an elevation they claim accounts for rising seas and risks of storm surge from hurricanes (units 3 and 4, which are now in their fifth decade of operation, sit twenty feet above sea level).

As part of its concurrent review process, the NRC evaluates emergency planning and other safety-related factors. Citizen activists have voiced frustration with the NRC's bifurcated approach to public health and safety concerns and lack of transparency. With climate change likely to produce stronger and more frequent hurricanes, the question of reactor safety has become more acute. The Southern Alliance for Clean Energy and other critics are troubled by the NRC's apparent disregard for its own critical analysis of the Fukushima nuclear disaster, despite disturbing parallels between the Japanese reactor design and Westinghouse's AP1000 nuclear design proposed for Turkey Point. Design-related safety concerns include untested

---

5. Southeast Florida Regional Climate Change Compact, "Region Responds to Changing Climate."

materials, an emergency cooling system that lacks the redundancy characteristic of other reactors, and risks associated with a meltdown of spent fuel, which is housed in open pools within a minimally secure structure adjacent to the containment building. Finally, despite the bankruptcy filing of Westinghouse in March 2017, Florida Power & Light still plans to use the same plant design and is searching for new manufacturing sources.

Florida Power & Light's organizational culture of silence is a further cause for concern, given how much the NRC relies on nuclear plant workers for information related to public health and safety. During the 1990s and into the 2000s, several Turkey Point employees filed complaints or quit in protest over serious safety violations. In the period from 2005 to 2011, the NRC received 160 anonymous complaints from Florida nuclear plant workers, far in excess of any other U.S. nuclear facilities.[6] Where speaking up should be the norm, a dangerous silence on issues of safety prevails.[7]

Turkey Point represents an all-too-familiar case of corporate power's ability to privatize profits and socialize costs while suppressing alternatives and capturing regulators.[8] So far, despite strong public support for energy conservation and alternative energy, the Sunshine State's powerful utilities and political establishment have ignored or blocked proposals for a clean-energy strategy driven by an energy-efficiency resource standard, a renewable portfolio standard, and an opening up of the energy market. Yet the politics of energy, development, and climate change in Florida is fluid, not frozen. While denial of the dangers posed by Big Energy remains strong, resistance is growing to the vulnerabilities imposed by vested interests.

6. Garcia-Roberts, "Next Nuclear Disaster."

7. Organizations with cultures of silence are hampered by poor decision-making, rigidity, and low employee morale. Where employees are disengaged, human errors and accidents are more likely. On cultures of silence, see Morrison and Milliken, "Organizational silence." Here it is worth recalling the root causes of the 2005 British Petroleum (BP) disaster in Texas City, Texas. Fifteen people were killed and 180 workers were injured in an explosion at BP's largest oil refinery. As a result, BP faced criminal prosecution and a multitude of lawsuits for failure to comply with federal environmental laws to prevent accidental releases of regulated substances. In 2007, the Chemical Safety and Hazard Investigation Board (CSB), a federal investigative agency, issued a report outlining the root causes of the accident: massive cost-cutting by corporate leaders; inadequate corporate oversight; outdated mechanical equipment; and a culture that discouraged the reporting of safety problems. Also in 2007, the Baker Panel, chaired by former Secretary of State James Baker and funded by BP in response to a recommendation by the CSB, found that workers at BP's US refineries were overworked, refineries were thinly staffed, and employees did not report accidents and safety concerns because they feared repercussions or judged that the company would not do anything about them. If it is true that "democracy dies in darkness" (as the *Washington Post* banner reminds us daily), it is also the case that workers die when silenced.

8. Klas, "Watchdog report."

As public trust erodes and demands for transparency grow, a space opens for local leaders, counter-experts, and concerned citizens to challenge the defenders of a nuclear-powered growth agenda. In April 2016 a Florida appeals court overturned the governor's green-lighting of the nuclear expansion project at Turkey Point. A month later Florida Power & Light informed the state's public service commission that it plans to postpone initial construction for Units 6 and 7 until 2020 even as it expects NRC approval for a new nuclear license in 2017.

Floridians are engaged in a high-stakes conflict over how risks are to be defined and distributed.[9] The various stakeholders must deal with an underlying issue, namely the irreducible uncertainty of a complex situation that defies all predictive models, cost-benefit calculations, scenario analyses, and other forecasting methods. Some recognize the limits of prediction and control, while others display confidence in the capacity of experts to deliver—within a range of probability—reasonably reliable projections. The latter position is exemplified by the NRC's rosy assessment of low and moderate impacts at Turkey Point, the former by critics who fear the current trajectory of techno-economic development in south Florida renders local communities and ecologies vulnerable to harm in ways that are not readily calculable, potentially catastrophic, and likely irreversible in the event of a major nuclear incident. These positions reflect opposite poles of the ongoing proactionary-precautionary debate.

In *The Black Swan*, Nassim Taleb reminds us of the "turkey problem," i.e., the age-old philosophical problem of induction. Imagine a turkey fed for several months in the run-up to Thanksgiving. As the weeks pass the turkey comes to expect—in the normal routine of things—a daily feeding; it is conditioned to anticipate the farmer will attend reliably to its needs. Yet on a Wednesday afternoon in late November, something altogether unexpected happens to the unwary bird. Taleb asks:

> How can we know the future, given knowledge of the past; or, more generally, how can we figure out properties of the (infinite) unknown based on the (finite) known? Think of the feeding again: What can a turkey learn about what is in store for

---

9. As Beck and others point out, the usual alignments are blurred in environmental risk conflicts. Note, for example, how the business camp in Florida is split, with some industries supporting the status quo while others join the opposition in calls for a shift to clean energy; while economic interest clearly colors risk perceptions, what "green" means is not always apparent to the players. Note as well how advocates of a nuclear ramp-up include some environmentalists, who argue that it's wiser to take on more nuclear risk, including the uncertainties of long-term nuclear waste disposal, in the effort to mitigate what are judged to be the greater and more proximate risks associated with climate change.

it tomorrow from the events of yesterday? A lot, perhaps, but certainly a little less than it thinks, and it is just that "little less" that may make all the difference.[10]

In world risk society, fewer people gobble up the assurances of safety fed to them by experts without first asking hard questions. In the case of Turkey Point, the more one learns the more questions appear. Flowing through the many specific issues—local hydrology, changing climate, reactor design, organizational culture, etc.—raised by nuclear energy expansion in south Florida is a practical and increasingly urgent question: How do we assess technologies in situations marked by complexity and scientific uncertainty, where impacts may be large (or catastrophic) and irreversible? The urgency of this question may not register, however, if on the day before Thanksgiving we sit down to read the paper and find ourselves captivated—yet again—by a Promethean-cornucopian vision of unending affluence.

An Evolutionary Ethics of Responsible Care for the Anthropocene

Philosopher Hans Jonas argued decades ago that the nuclear, chemical, genetic, and ecological threats generated by techno-economic progress demanded a new ethics of responsibility. The coming of world risk society renders traditional ethics less relevant, since earlier assumptions regarding the limited temporal and spatial impact of human actions no longer hold. Under qualitatively new conditions, inter-generational justice and the survival of humanity become paramount ethical concerns: "A kind of metaphysical responsibility beyond self-interest has devolved on us with the magnitude of our powers relative to this tenuous film of life, that is, since man has become dangerous not only to himself but to the whole biosphere."[11] With the advent of the Anthropocene, an abiding concern for the well-being of one's offspring is, for Jonas, the paradigmatic instance of responsibility for those to come.

---

10. Taleb, *Black Swan*, 40. Beck makes the same point in the context of what he refers to as the "organized irresponsibility" of modern industrial risk-management: "The unknowable risks of the risk models hide behind the façade of controllability. Since modern forms of risk management for the most part maximize mathematical precision, they systematically underestimate unforeseen and improbable, but not therefore impossible, occurrences, as regards both their frequency and the extent of the damage they cause. This apparently slight difference between 'improbable' and 'impossible' makes a world of difference" (*World at Risk*, 130).

11. Jonas, *Imperative of Responsibility*, 136.

In light of Turkey Point and other troubling cases, it is an open question whether industrial societies possess the legal, political, and cultural resources required to meet the challenges of a risky, runaway world. As Rosa argues, hyperacceleration not only aggravates the problem of desynchronization in a complex, functionally differentiated society, it also throws a cloud over collective action and democracy itself. Technology and economics race ahead of democratic decision-making and policy implementation, resulting in a reactive, situational politics devoid of long-range vision. For his part, Jonas worried that technological hubris might silence or ignore wiser counsel informed by the "heuristics of fear," which he hoped would generate a deep sense of caution and the collective will to limit technology for the sake of preserving the "tenuous film of life." Yet Jonas was skeptical toward democratic institutions, mired as they are in interest-group conflicts and structurally oriented to compromise solutions utterly inadequate to the threats. He wondered aloud whether authoritarian government might be necessary to save humanity from itself if freedom became untethered from a sacred responsibility to preserve life above all else. Such eco-fascist solutions must be rejected (and Jonas himself did reject this option), yet the critique of liberalism by Jonas, William Ophuls,[12] and others has merit in light of industrial modernity's "self-annihilating progress" (Beck) and a late-modern politics rendered rudderless by hyperacceleration. Liberal polities—still caught within the nation-state framework and committed, philosophically and practically, to individualism—face a legitimation crisis that shows few signs of resolution.

Along the path pioneered by Jonas (but with a greater appreciation for what technology and democracy might accomplish than he allowed), I propose an *evolutionary ethics of responsible care* capable of informing and inspiring a politics of engagement and time. The proposed moral framework has four salient features: (1) it is a global or common ethic informed by the new cosmology and grounded in the universal capacity to care,[13] and thus available in principle to all persons; (2) for persons of faith, it is consistent with an eco-theological vision of the human encounter with unfolding plenitude in universal history; (3) it is at the same time a place-based ethic that affirms enduring attachments to local communities and landscapes, cares about their distinctive character and unique ensemble of biographies and histories, and is thus motivated to develop community-based strategies

---

12. Ophuls, *Requiem for Modern Politics*.

13. Weld, *Ethics of Care*. Recent research in ethology and evolutionary biology, neuroscience, and behavioral economics supports the claim, first made by Darwin, that care is an evolved social instinct shared by all humans and their evolutionary kin; see de Waal, *Age of Empathy*.

for adaptive fitness as climate and other conditions change; and (4) it provides actionable guidance across a range of settings and is supportive of participatory technology assessment and other deliberative-democratic practices that require a political commitment to modulating the pace of technological development when necessary, e.g., through a judicious application of the precautionary principle. Since every ethic implies a worldview (and vice versa), I begin with Kevin Kelly's intriguing thesis that, over the long haul, technology exhibits the same exotropic (or negentropic) tendencies as nature.

Rivers of Mutations

In *What Technology Wants*, Kelly folds a defense of the convergent evolution thesis into a larger claim about the directionality of nature and technology. From the Big Bang onward, an evolving universe continues to generate tendencies that swim against the cosmic tide of entropy. Contra Gould et al., nature is not wandering aimlessly in its evolutionary unfolding, with life and mind mere accidents along the way; rather, the cosmos exhibits a persistent, if uneven and sometimes halting, tendency to evolve in certain directions.[14] Specifically, Kelly identifies thirteen exotropic tendencies: the universe tends over time toward increasing efficiency, opportunity, emergence, complexity, diversity, specialization, ubiquity, freedom, mutualism, beauty, sentience, structure, and evolvability.[15]

What makes Kelly's soft-teleological reading of evolution interesting is the claim that all thirteen exotropic tendencies can be seen in both natural evolution and technological development, or what he calls the "technium"—the accumulating stock of cultural inventions used to build up civilizations over thousands of years. The invention of musical horns, for instance, occurred in more than one locale, and over the centuries the family of horns (and related forms of music) has grown more diverse, complex, specialized, and so on. By the time John Coltrane receives an alto sax for his seventeenth birthday, the possibilities for expressions of beauty with a wide variety of horns have expanded greatly—and will expand still further as Trane sets a new jazz standard, the next generation arrives, and new ways to design and play horns are invented.

---

14. Among evolutionary biologists, the convergent evolution thesis is defended by Morris, *Runes of Evolution*.

15. Kelly, *What Technology Wants*. While Kelly's outlook is similar to that of Teilhard de Chardin, he does not appear familiar with the Jesuit paleontologist's work.

A major difference between natural evolution and social history, of course, is that human societies generate and guide technological development (including technologies for shaping our own humanity through genomics) instead of being subject to the slow, trial-and-error process of natural selection. Allied to this difference is the exponential pace of technological change. As Kelly observes,

> It took Sapiens several million years to evolve from an apelike ancestor. During that transition to humanity, our DNA changed a few million bits. So the natural rate of biological evolution in humans, in terms of information accumulation, is about one bit per year. Now, after almost four billion years of bit-by-bit biological evolution, we have unleashed a new type of evolution, one that creates rivers of mutations using language, writing, printing, and tools—what we call technology. Compared to the one bit per year we made as apes, we are adding 400 exabytes of new information to the technium each year, so the rate of our technological evolution is a billion billion times as fast as the evolution of DNA.[16]

Three important ethical implications follow from Kelly's worldview. First, despite differences in the pace, mechanisms, and dynamics at work, we can affirm the unity and continuity of natural and sociocultural processes of evolution. As Kelly puts it, "a single thread of self-generation ties the cosmos, the bios, and the technos together into one creation." In other words, the new cosmology offers a divided, conflict-ridden humanity a *common story* that affirms a dynamic unity-within-diversity and sense of shared destiny at the heart of an evolutionary ethics of responsible care. Second, the directionality we discern in evolution is open-ended, not preordained, and hence humanity has an active, co-creative role to play in shaping the future. With the fast-accelerating development of the technium and entry into the Anthropocene, humans now must assume greater responsibility for the fate of the planet. Third, the exotropic tendencies at work in both nature and technology possess immense normative significance. We can and should assess both natural processes and technological developments based on the degree to which—on balance—they promote or hinder these tendencies toward the creation of greater value and beauty.[17] In sum, a global ethic of

---

16. Ibid., 334.

17. Attentive readers will note how Kelly implicitly derives an "ought" from the "is" of exotropic tendencies, and thus commits the naturalistic fallacy (Hume). In a postmodern context, where the search for direction in the movement of life proceeds without the compass of classical metaphysics and natural law, Christians must nonetheless risk articulating as coherent and comprehensive a worldview as possible. Kelly's

responsible care arises from and is informed by the evolutionary epic, an unfolding drama with tendencies we can glimpse and be guided by as we envision and enact the next scene.

Informed by Kelly's worldview as well as Aldo Leopold's land ethic, we can state the basic maxim of an evolutionary ethics of responsible care as follows: A thing is right when it tends to promote the planetary bent toward increasing diversity, complexity, structure, sentience, freedom, and beauty; it is wrong when it tends otherwise. Alternately and more abstractly, we might say: A thing is right when it tends to promote the exotropic tendencies of the universe; it is wrong when it tends otherwise.[18] Needless to say, this maxim leaves many questions unanswered, yet its potential value in guiding ethical-political deliberation is considerable, in part because it provides orientation and a broad, inclusive sense of direction in the movement of life.

---

reading of evolution is compatible with a process-oriented eco-theological vision that affirms the ontological goodness of creation, construed in dynamic fashion as the ongoing realization of greater value and beauty in an evolving cosmos. As Gamwell argues, the ground for any moral claim is the divine good (as understood within a process-metaphysical framework); see Gamwell, *Divine Good*. In reply to the hermeneutical suspicion of ecofeminists, social ecologists, and other critical-theoretical perspectives, I suggest an eco-theological vision informed by an evolutionary-ecological worldview is less likely to reinstate oppressive binaries or reinforce unjust power structures. This claim is open to contestation, of course. For example, some critical theorists will find an anthropocentric bias lurking within any grand narrative of evolution that recognizes the transformative power of human cultures in altering earth's biosphere.

18. In *A Sand County Almanac* (1949), Leopold famously states: "A thing is right when it tends to preserve the integrity, stability and beauty of the biotic community. It is wrong when it tends otherwise." Leopold's dictum has been criticized for its presumably static view of the biotic community (or simply, the land). However, Callicott argues that Leopold's mature thought makes room for autopoiesis and change—including anthropogenic alterations that can enhance the land—in the evolution of ecosystems; see Callicott, *Beyond the Land Ethic*, chap. 17. In light of Callicott's reading, I venture to revise Leopold's classic statement in two ways, first by shifting emphasis from "preserving" to "promoting," and second by altering and expanding the valued characteristics from three (i.e., integrity, stability, and beauty) to six (where the six enumerated imply all of the exotropic tendencies Kelly identifies). Doubtless some will object to the omission of integrity and stability in any revised statement that claims faithfulness to Leopold. This concern is valid—to a point. In reply, I suggest any sensible and sensitive use of Kelly's set of exotropic tendencies (in combination with his criteria of conviviality discussed below) for evaluating the potential impact of techno-economic development will consider fully the need to respect ecological integrity and stability. Here insights from ongoing research in sustainability science will prove invaluable.

Becoming More

From an eco-theological perspective, persons of faith can affirm Kelly's reading of the new cosmology and the evolutionary ethics of responsible care that flows from it. Through evolution a divine creativity flows unceasingly, inviting all things to become *more*. As John Hart puts it, "evolution is an autopoietic process of biotic creation as the Creator-initiated Singularity (the explosive origin of an inflating universe) unfolds in dynamic creative freedom within physical, chemical, and biological parameters; the Spirit continues to create, now indirectly, as cosmic and planetary entities and processes interact and integrate."[19] Humanity's vocation, we have said, is the convivial co-creation of greater value and beauty in an evolving cosmos. We are actors caught up in the grand drama of creation-in-process, yet we also have a hand in writing the script. Authentic human development is realized in and through creative, responsible, and caring participation in this ongoing adventure, a participation in which we are drawn continuously by an alluring Spirit into a more profound, intimate, and mysterious encounter with unfolding plenitude in universal history. As John Haught eloquently states, the universe "continues to be drawn toward an unpredictable and open future by the attractive power of a God who creates the world by offering it new possibilities for becoming more—opportunities for more intense and valuable modes of being . . . The ultimate explanation of evolution is the coming (or advent) of God into the world from out of an endlessly expansive future."[20] Every human expression of caring, creativity, and moral responsibility occurs within this dramatic, adventurous context. All the while and at every turn we are drawn forward by the promise of more to come, and of becoming *more*. Moreover, as noted above, the broad trajectory of this "more" can be glimpsed through the new cosmology. Indeed, creation itself is the primary revelation, as Thomas Berry says.[21]

In This Place

Many particular stories fold into the larger, unfolding cosmic narrative. Even with global connectivity and the disembodied Second Life we create online, our lives still unfold largely within specific *places* and the thicket of relationships and memories tied to them.[22] We remain creatures embedded within

---

19. Hart, *Cosmic Commons*, 4.
20. Haught, *Making Sense of Evolution*, 138.
21. Berry, *Sacred Universe*.
22. For a philosophical defense of place as integral to human flourishing, see

local communities and landscapes, formed by our emotional entwinements with and loyalties to the people upstairs, across the street, down the block, in the neighborhood, across town, around the region, and within the watershed. We care for the cats, canaries, and other critters (domesticated and wild) inhabiting our home territory, counting them our own. For most of us, a global ethic and grand cosmological epic becomes grounded in reality when it lands in our backyard, as it were, the unique places and cherished relationships where our responsible care is rooted even as wider horizons beckon and more inclusive ethical commitments make claims upon us.[23]

Hence in a risky, runaway world the places that local and indigenous communities inhabit and call home are privileged locales for putting an evolutionary ethics of responsible care into practice.[24] As climate and other conditions change, the demands of adaptive fitness will require of prophetic-utopian imaginative thought a deep respect for local/indigenous agency and traditions (without freeze-framing their authenticity).[25] Any place-based, ethical-practical response to the local fallout of climate change must avoid reinscribing patterns of discrimination and social exclusion. A shared awareness of the need for adaptive changes can serve as a point of departure for collaborative exercises in visioning and problem solving in the face of unsettling shifts in the environment.

Cannovo, *Working Landscape,* chap. 1.

23. A rich relatedness to place, community, and tradition can be stifling, of course. It is difficult to maintain a creative tension between tradition and innovation, roots and wings, or what Cannovo (following Iris Marion Young) calls preservation and founding. As Keller puts it so well, "The relationships composing our memory, body, family, community, ecology, and world form the material, largely unconscious, of our ongoing genesis. And these very relationships can close in on our spirit, close us down, usually in the name of morality and tradition . . . Nonetheless, the ambiguous matrix of our relationships also constitutes the material through which spirit *matters*. In resistance to structures that have formed and deformed relationship itself, but only in and through these relations, can 'I live out what I am'" (*On the Mystery*, 106).

24. Cf. Norton, *Sustainability,* 334–38; and Shutkin, *Land That Could Be.* See also Northcott's Christian argument for a place-based ethics and politics of sustainable community in *Place,* chaps. 7–8.

25. Where heterogeneous communities grapple with large and disruptive changes on the horizon, both liberal guilt and the real danger of continuing violence against minority/indigenous groups must be kept in check. Here it seems helpful to recall what Iain Chambers has said about making a fetish of cultural authenticity: "To talk of authenticity has invariably involved referring to tradition as an element of closure and conservation, as though peoples and cultures existed outside the languages of time. It is to capture them with the anthropological gaze, where they are kept in isolation and at 'critical distance' as though they do not experience movement, transformation" (*Migrancy, Culture, Agency*, 21).

One good example of multi-stakeholder collaboration is the four-county compact to address climate change in south Florida mentioned above, another the creative effort across the pond of two artists, Nelson and Helen Harrison:

> Their *Greenhouse Britain* (2007) is built around scenarios concerning the effects of rising sea levels on the Mersey Estuary and the Lea Valley in the United Kingdom. By working over an extended time with local people, planners, scientists and policymakers who live and work in these areas, the Harrisons produce near-term science-fiction scenarios that exploit local knowledge of connectedness to imagine how communities will change their ways of living.[26]

An inclusive, collaborative project, *Greenhouse Britain* (subtitled "Losing Ground, Gaining Wisdom") enables coastal residents to voice their fears while also offering them a way to envision the adjacent possible of sustainable community amidst uncertainty and acutely felt vulnerability.

Convivial Technologies

Finally, Kelly's outlook on nature and technology provides a framework for technology assessment consistent with an evolutionary ethics of responsible care. The thirteen exotropic tendencies, he suggests, are useful as guidelines for evaluating a technology's trajectory:

> This list of exotropic trends can serve as a sort of checklist to help us evaluate new technologies and predict their development. It can guide us in guiding them . . . We can compare competing technologies to see which one favors more of these exotropic qualities. Does it open up diversity or close it down? Does it bank on increasing opportunity or assume they wither? Is it moving toward embedded sentience or ignoring it? Does it blossom in ubiquity or collapse under it?[27]

Kelly also proposes six criteria—cooperation, transparency, decentralization, flexibility, redundancy, and efficiency—for assessing "convivial manifestations" of technology. In practice, an evolutionary ethics of responsible care informed by these guidelines and criteria requires constant engagement with developing technologies and perpetual vigilance in monitoring their impacts. Kelly's proactionary approach calls for anticipation, continual

26. Groves, "Living in Uncertainty," 123.
27. Kelly, *What Technology Wants*, 271.

assessment, prioritization of risks (including natural ones), rapid correction of harm, and redirection rather than prohibition of technologies.

Jonas, we observed earlier, considers parental concern for the well-being of offspring as the paradigm for responsibility to future generations. Kelly takes up the parent-child relationship as a useful analogy for describing how humanity can manage technology well. Good technologists, like attentive parents, will be vigilant and pick up quickly on a technology's "anti-social" behavior. A deft redesign or thoughtful redeployment (or both) of the problematic technology might follow, in ways analogous to the skillful redirecting of a wayward child.

How might Kelly's guidelines and criteria be applied to the nuclear-versus-renewable energy debate in Florida (and elsewhere)? At a glance it seems apparent that Florida Power & Light's choice of nuclear (and natural gas)[28] fares poorly in comparison with an alternative mix of renewable sources (i.e., a combination of solar, wind, and possibly gulfstream power) when assessed on the basis of the extent to which each option promotes exotropic tendencies over time and meets the six criteria for convivial expressions of technology. Deploying the latter to assess energy options in Florida, it appears clear that a smart mix of renewable sources (led by solar, a superabundant resource in the region) is preferable on all six counts: it does a better job of promoting collaboration between people and institutions (cooperation); its workings are intelligible to non-experts, and there is no asymmetrical advantage of knowledge to some of its users (transparency); its ownership, production, and control are distributed (decentralization); it is easy for users to modify, adapt, improve, or inspect its core, and individuals can choose to use it or not (flexibility); it is not the only solution, not a monopoly, but one of several options (redundancy); and it minimizes impact on ecosystems, has high efficiency for energy and materials, and is easy to reuse (efficiency). In sum, Floridians can live more convivially with a mix of renewable technologies than with nuclear reactors, the cores of which are always too hot to handle and which may undergo meltdowns that even the best of professionals cannot control.

---

28. Note that Florida Power & Light (FPL) also operates generators fueled by natural gas at Turkey Point. In recent years the utility has formed joint ventures with energy firms in the Midwest that are engaged in natural-gas extraction using the method of hydraulic fracturing. FPL's sourcing of "fracked" natural gas will be funded by ratepayers, not investors; see Klas, "FPL customers."

Between Recklessness and Relinquishment, the Creative Middle

While Kelly recognizes the risk of major, irreparable harms from technology deployments, he sides with philosopher Max More and other supporters of a proactionary approach that emphasizes the freedom to innovate and holds that the burden of proof falls on those proposing restrictive measures.[29] On this view, unintended consequences are to be expected and managed accordingly. In Kelly's historical reading of the growing technium, the benefits of new technologies tend on average to outweigh the (inevitable) harms. Moreover, a strategy of relinquishment—e.g., prohibiting development of synthetic biology, nanotech, robotics/AI, and other emerging technologies—would have the ironic and unwelcome result of generating even more risk. Relinquishment in practice would require strong government control, which in turn would inhibit open innovation and drive research underground.

We may ask, then, when is precaution required? Advocates of the precautionary principle correctly reply: whenever a potentially large scale and irreversible threat to human and environmental health appears. Precautionary measures are advisable in situations with certain characteristics: scientific uncertainty, potential largescale impact and cumulative harm, potential latent effects, potential irreversibility, and imperceptibility.[30] Especially in cases where institutional capacities to redirect a technology and mitigate its harmful effects may be inadequate, the optimism and daring of proactionaries edges too easily into recklessness. Hence in law and public policy, a proactionary approach needs to be balanced with judicious applications of the precautionary principle, given the kinds of dangers we face in world risk society and the perennial problem of black swan blindness.

---

29. Transhumanists have articulated versions of a proactionary principle. More states it thus: "People's freedom to innovate technologically is highly valuable, even critical, to humanity. This implies a range of responsibilities for those considering whether and how to develop, deploy, or restrict new technologies. Assess risks and opportunities using an objective, open, and comprehensive, yet simple decision process based on science rather than collective emotional reactions. Account for the costs of restrictions and lost opportunities as fully as direct effects. Favor measures that are proportionate to the probability and magnitude of impacts, and that have the highest payoff relative to their costs. Give a high priority to people's freedom to learn, innovate, and advance" ("Proactionary Principle").

30. Among various formulations of the Precautionary Principle, the Barnier Law (France, 1995) is a typical example: "The lack of certainty ... must not delay the adoption of effective and proportionate measures that aim to prevent a risk of serious and irreversible damage to the environment."

Finding the right balance is a difficult challenge. Rather than taking sides in the proactionary-precautionary debate, I suggest we adopt a creative-middle approach that can help us to navigate between the Scylla of recklessness and the Charybdis of relinquishment in technology development. Proactionaries are prone to one extreme, precautionaries to the other. While society cannot afford to play with the fire of high-risk technologies, it also cannot afford to deep-freeze innovation processes that yield promising new inventions. That said, we need more than a golden-mean approach. As the proactionary-precautionary debate plays out at several levels—from controversies over specific technologies to deeper paradigm conflicts—a creative-middle approach will not seek out compromise, but rather will seek to engage all stakeholders in a generative dialogue and collaborative-learning process that leads to new insights, breakthroughs in understanding, novel experiments, and discovery of alternative pathways, none of which the parties involved could produce on their own.[31] In situations where dialogue and collaborative learning fails to generate a new way forward, or where enabling conditions for productive dialogue and collaboration are absent, it may be necessary to stake out a principled yet flexible middle position. In such cases, what constitutes a responsible stance will be subject to deliberation and debate, and in principle it should be open to revision and change as conditions evolve, new evidence appears, and alternatives are developed. Finally, in these less-than-ideal (and more likely) situations, ambiguity and political conflict appear inescapable, given high levels of scientific uncertainty and complexity, asymmetries of information and power, the limits of foresight, ideological bias, and other factors in play.

The conceptual and practical challenges involved in taking a creative-middle approach may be illustrated with reference to the ongoing debate over synthetic biology. Consider the work of the US Presidential Commission for the Study of Bioethical Issues. Its 2010 report, titled "New Directions: The Ethics of Synthetic Biology and Emerging Technologies," identifies five principles of assessment: (1) *public beneficence*—adopt "societal perspective" in weighing risks and benefits; (2) *responsible stewardship*—exercise "prudent vigilance" and avoid extremes; (3) *intellectual freedom and responsibility*—maintain oversight governed by a corollary principle: regulatory parsimony; (4) *democratic deliberation*—conduct a "careful ongoing review of the science and its applications"; and (5) *justice and fairness*—seek a just distribution of benefits, burdens, and risks. This set of principles may be read as an attempt to mediate between proactionary and precautionary viewpoints. For instance, the former is reflected in principle #3, while #4

---

31. This is the type of generative dialogue that Pope Francis calls for in *Laudato si'*.

and #5 are hallmarks of the latter. Both stances are committed to #1 and #2; however, what "prudent vigilance" requires in practice is a major topic of debate. In 2010 the commission took the position that no regulation of synthetic biology research was required at present. They urged regulators and researchers to exercise prudent vigilance by collaborating on development of methods, tools, and protocols for testing and monitoring.[32]

Did the commission's work reflect a responsible middle position, or did its principles and no-regulation-yet recommendation amount to little more than a political compromise among commission members? Green-communitarian groups suspected the latter and thus rejected the commission's position. They called instead for a strong precautionary approach, including a moratorium on the release and commercial use of synthetic organisms until a robust legal framework and strong regulatory infrastructure for protecting human and environmental health could be put in place.[33] For these groups, located at the margins of power, the need for precaution and protection arises from concerns over bio-colonialism and reflects a commitment to an alternative model of the humanity-nature relationship.

Green communitarians, we noted in chapter 2, call for epistemic humility and reverence toward Gaia. On this view, nature is not more complex than we know, it is more complex than we *can* know. This is particularly true with modern technological interventions, the consequences of which are not simply unintended but irreducibly indeterminate as well as irreversible

---

32. Since the commission's 2010 study, initial steps toward regulating commercial synthetic biology products has begun, but the novel character of many technologies has confounded federal laws and regulatory agencies set up during the industrial era. Issues of jurisdiction, poor inter-agency communication, and lack of technological literacy have been identified. A gap analysis and recommendations for modernizing the regulatory infrastructure are presented in a Wilson Center white paper, "DNA of the U.S. Regulatory System." From the perspective of Beck's world risk society theory, the situation presents another example of "organized irresponsibility" in the making. And viewed in light of Rosa's social-acceleration theory, the temporal gap between leading research and lagging regulation could not be wider.

33. Friends of the Earth, "Principles for Oversight." This document has been signed by over a hundred civil-society groups from dozens of countries. Seven regulative principles are proposed, beginning with a well-known version of the precautionary principle, the "Wingspread Consensus Statement" (January 26, 1998): "When an activity raises threats of harm to human health or the environment, precautionary measures should be taken even if some cause and effect relationships are not fully established scientifically. In this context the proponent of an activity, rather than the public, should bear the burden of proof. The process of applying the Precautionary Principle must be open, informed and democratic and must include potentially affected parties. It must also involve an examination of the full range of alternatives, including no action." http://sehn.org/wingspread-conference-on-the-precautionary-principle/.

in many cases.[34] Adopting a Jonasian heuristics of fear, greens are suspicious of the "engineer's gaze," which reduces nature to solvable problems; the system's past failures make them wary of technological hubris. Moreover, the "re-making Eden" project smacks them as complicit with a bio-colonial conquest narrative; they point out that "most commercial interest in synthetic biology is focused on enabling a new 'biomass-based economy' in which any type of plant matter can be used as feedstock for tailored synthetic microbes to transform into high value commercial products—anything from fuels to plastics to industrial chemicals."[35] They fear a shift to large scale biomass production in the global South will concentrate land ownership in fewer hands, reduce biodiversity further, and endanger rural livelihoods. At stake is the basic question: Which strategic vision and policy narrative for sustainable development?

Synthetic biologists reject these charges, meanwhile harboring their own suspicions that "radical opponents" misunderstand the science and fail to appreciate its potential benefits. They view themselves as a techno-progressive vanguard boldly exploring a new frontier.[36] It appears, then, that different stances on scientific practice, practical rationality, and public policy stem from rival epistemologies, competing historical narratives, and two very different imaginations, one arcadian and the other enlightened technocratic. A political battle over research agendas, regulations, and sustainable-development strategies may reflect a deeper paradigm conflict.

Faced with this type of stand-off, efforts to establish a creative-middle stance may seem impossible or naïve. What responsible care seems to require is resolute commitment to either one position/paradigm or the other. However, this type of either-or framing harbors its own dangers, since it may legitimize an intractable conflict that precludes the possibility of generative dialogue, social learning, and institutional reinvention. Thus a creative-middle approach must first try to discern whether the situation as framed is indeed a genuine paradigm conflict or only apparently so. If considered judgment deems it a deep-seated conflict of paradigms, then a further judgment needs to be made on where one stands and reasons given to justify the commitment. (Yet even here, those navigating the creative middle remain committed to dialogue.) But if the framing is judged open to question, then a creative-middle approach will focus on possible reframings and alternatives by promoting further research, shared inquiry, and frank discussion among stakeholders.

34. See, e.g., Huesemann, *Techno-Fix*, chaps. 1–2.
35. Friends of the Earth, "Principles for Oversight," 11.
36. See, e.g., Venter, *Life at Speed of Light*.

Using Lonergan's Dialectics as a Navigation Tool

These deliberative considerations suggest the need for further conceptual resources in mediating the proactionary-precautionary debate and discerning viable pathways forward. Along with the practical guidance offered by an evolutionary ethics of responsible care, Bernard Lonergan's heuristic account of the dialectics of community and culture can serve as a navigation tool or compass of sorts as we search for direction and attempt to steer away from dangerous trajectories in technology development. As an alternative to the proactionary optimism of Kelly, More, and other libertarian thinkers, Lonergan provides us with a theological-ethical interpretation of the dynamics of technological, social, and cultural developments within history that takes into account human sinfulness, or what he describes as the distorting impact of various biases. On this reading, the history of technology is part of a larger, unfolding drama of human progress, decline, and redemption.

According to Lonergan, the development of societies within history involves a constructive unfolding of two opposed but linked principles, one a grounding principle of limitation, the other an expansive principle of transcendence. These opposites-in-tension constitute an integral, unfolding dialectic of contraries that Lonergan calls the "dialectic of community."[37] At the pole of limitation, spontaneous intersubjectivity (fellow-feeling) grounds and sustains human community (and, we'd add, our innate capacity for biophilia grounds and sustains our felt connection to the natural world).[38] At the pole of transcendence, the continual exercise of practical intelligence generates larger and more complex technical, economic, and political systems (or what Kelly refers to as the "technium"). Societal progress occurs when this integral dialectic unfolds constructively, i.e., when a dynamic balance or creative tension between the two poles is maintained as individuals and communities solve the problems of living through intelligent, responsible action.

The conflicting demands of group feeling and intelligent practicality are mediated by culture, which generates a more or less coherent set of constitutive meanings and values that inform and guide the practices and institutions of society. However, individual and group biases may distort or deny cultural meanings and values, as when patriotism and security morph dangerously into chauvinism and a police state. Such distortions register

---

37. Lonergan, *Insight*, 211–18. The dialectic of community is parallel in its formal structure to the dialectic of the subject discussed in chap. 6 above.

38. Kellert and Wilson, *Biophilia Hypothesis*.

within the dialectic of community as an unhealthy shift or displacement toward one pole or the other.

Absent the limited solidarity arising from a response to the threatening Other, many of the "imagined communities" (Anderson) making up the modern nation-state system have struggled to establish and maintain a unifying sense of shared meanings and values. Today, most elites in advanced-industrial societies settle for a vague utilitarian commitment to the general welfare based on an early-modern belief in techno-economic progress, while dissidents of various stripes reject liberal modernity and what they see as its pretentious claims of progress. Consequently, contemporary societies are prone to the "idolatry of technique" (Ellul), on the one hand, and tempted by an atavistic urge to return to an allegedly "simpler time," on the other. With the former, society moves too far (and too *quickly*, we would add) toward the pole of transcendence. In this case, instrumental reason prevails and technological determinism takes hold, resulting in the fragmentation and destruction of both social and ecological communities. The latter temptation yields an unhealthy move too far toward the pole of limitation; here nostalgia retards healthy growth and change, resentment dampens creativity, or regression to collectivism ends up stifling individuality and critical rationality. Either way, societal decline stems from a distorted dialectic.

A paradigmatic case of modern decline occurs wherever aggressive, military-industrial expansion runs roughshod over social and natural ecologies. The state-sponsored application of brute-force technology disrupts life-worlds, erases local cultures, and disturbs the rhythms of nature—always in the name of progress and security. Among many examples, we may recall the Argentine *criollo* government's genocidal conquest of Patagonia in the 1870s, Mao's war on nature and destruction of peasant communities during the 1950s and 1960s,[39] the forced relocation of Miskito Indians from coastal homelands to high-altitude communal coffee plantations by Nicaragua's Sandinistas in the 1980s, the US Navy's current, large scale assault on whales and dolphins through weapons and sonar testing, and the Indonesian government's continuing collusion with corporations in clear-cutting peat forests and replacing them with palm-oil plantations. This characteristic form of modern decline represents what Lonergan scholar Robert Doran refers to as a "mechanomorphic distortion" of the dialectic of community, i.e., a drastically destructive shift toward the pole of transcendence and consequent devaluing and atrophication of innate human sociality and biophilia. In the modern era, the temporal dimension of this distortion typically is registered

39. Shapiro, *Mao's War Against Nature*.

by severe desynchronization of subsystems within society, and by a broader temporal disjuncture between high-speed society and the slower rhythms and processes of nature. Today we see "blessed unrest" (Hawken) within world risk society precisely because millions of people sense deeply that a dangerous imbalance—or what we are calling a distorted dialectic of community—now threatens the health and survival of both society and nature.

The dialectic of culture, itself an integral dialectic of contraries with opposed but linked principles existing in creative tension, involves three basic dimensions of constitutive meaning in society.[40] According to Doran, in human history the collective search for truth has disclosed cosmological, anthropological, and soteriological forms of constitutive meaning. The search for *cosmological truth* is driven by a profound need for a sense of cosmic order and for harmony between humanity and nature. Doran says that "cosmological truth is the discovery that direction in the movement of life lies in a harmony between human decisions and actions, on the one hand, and the rhythms and processes of nature, on the other hand, that is, in a synchronicity of culture and nature."[41] In pre-modern societies, religious mythologies functioned as carriers of constitutive meaning; through ritual reenactment a society constituted its life-world and renewed itself cyclically; social order was dependent on establishing and maintaining sacred cosmic order.

During the axial period (Jaspers), *anthropological truth* is discovered as the search for meaning and purpose in life reaches beyond the (mythically constituted) cosmos to a transcendent reality. The axial breakthrough during this time (800–300 BCE) amounts to "a shift in the orientation of consciousness to the source of cosmic and social order," i.e., to the discovery of a world-transcendent Reality or Absolute. Within a cosmological horizon, primal mythic consciousness sacralizes the powers of nature; the "gods" are identified with the rhythms and processes of the cosmos, and so human attunement with these "divine powers" (e.g., the sun) is paramount. By contrast, mystical-sapiential consciousness transcends the realm of intracosmic powers in search of the one God "beyond the gods."

Whereas the anthropological truth discovered by mystics and sages is humanity's striving toward the divine, *soteriological truth* denotes God's initiative to redeem humanity and bring all creation to its fullness (Eph 1:10). This saving truth, Doran notes, "is incarnate in a pattern that is experienced

---

40. Doran, *Theology and Dialectics of History*. As Doran puts it, "The culture that conditions an integral dialectic of community is . . . itself dialectical" (503). Doran's account of the dialectic of culture expands upon Lonergan's dialectics of the subject and community.

41. Ibid., 216.

even by mystic philosophy as startlingly strange, the pattern of suffering servanthood assumed by the world-transcendent measure of integrity become human flesh."[42] For Christians, soteriological truth's most powerful—indeed singular—expression is in the Christ event (and specifically in what Lonergan calls the "mysterious yet just law of the cross"). Soteriological truth finds continuing expression in history through the redemptive praxis of the saints. Divine grace redeems humanity from its sinful biases and establishes a friendship between God and the human community.

The creative tension existing between cosmological and anthropological truth is maintained in and through the realization of soteriological constitutive meaning. This meaning is *incarnate*, i.e., embodied in exemplary lives committed to redemptive praxis, which reverses decline by healing or overcoming biases and by opening up and pursuing hitherto neglected or undervalued possibilities for creating a different and better world. The most intense, crystallized expressions of incarnate meaning—e.g., the lives of St. Francis, Harriet Tubman, Mahatma Gandhi, Dorothy Day, Dr. King, Nelson Mandela, Oscar Romero, Dorothy Stang, and many other transformed, transformational leaders—disclose an alternative future and bring it to partial realization. Through their missions, they reverse decline and promote genuine progress. Today, we see dynamic, visionary leaders in many walks of life—e.g., Paul Farmer in the field of public health, Dee Hock in the business world—articulating and acting on visions of the adjacent possible that are informed (implicitly) by an acute, penetrating sense of the integral dialectics of community and culture. These inspired, inspiring lives show what the convivial co-creation of greater value and beauty looks like in practice.

Within the dialectic of culture, cosmological truth (roots) functions as a principle of limitation, and anthropological truth (wings) as a principle of transcendence. Taken singly, the dimensions of cosmological and anthropological truth remain incomplete as sources of meaning and value for shaping the practices and institutions of society. An integral dialectic of culture involves a constructive unfolding of these opposites-in-tension, while a distorted dialectic results from the deleterious effects of various biases. With the modern "mechanomorphic distortion" of the dialectic of community now at a critical stage, we find considerable energy and creativity devoted to rebalancing this dialectic through a culturally mediated "return to nature and community," artful expressions of cosmological truth (e.g., Cameron's film *Avatar*, Goldsworthy's site-specific sculpture and land art, Quinn's philosophical novel *Ishmael*), and various forms of environmental activism, including a vigorous defense of indigenous cultures. This rebalancing also

---

42. Ibid., 487.

is reflected in papal social teaching: a "pope of technology" (Pius XII) has been followed by a "light green" pope (Paul VI), two green popes (John Paul II and Benedict XVI), and now by Francis, the first "dark green" pope.[43]

To sum up, Lonergan's dialectics of community and culture helps us to maintain an incisive, creative-middle stance in dialogue with techno-progressives and bio-conservatives, since it addresses the question of technology from a more comprehensive viewpoint that honors the insights and values—and sees the blind spots—of both sides. Lonergan's heuristic framework is useful as a navigation tool or compass that enables us to detect and correct tendencies to veer dangerously off course. On the one hand, we should be prepared to critique technological determinism under whatever false-utopian guise it may appear. Jonas pointed out the follies of a Marxian technological utopian project. Today, we need to expose the hubris and ethical vacuity of some in Silicon Valley who breezily embrace a technological Singularity, or the duplicity of powerful bio-technology interests who claim that society has no choice but to adopt and scale up the latest agro-industrial product on offer. On the other hand, we must resist the fear propagated by Chicken Littles, including voices calling for policies of technological relinquishment and overly strict applications of the precautionary principle.

The Creative Middle and Reforms of and within Technology

A creative-middle approach will be *creative* (and constructive) to the extent that it continually seeks out further insights and puts forward viable, promising alternatives for developing convivial manifestations of technology consistent with an evolutionary ethics of responsible care. Adopting a principled yet flexible creative-middle stance can play a crucial role in exposing the limits and problems of current techno-economic regimes, while also promoting the development of alternative methods and models for guiding technology's trajectories.[44] The ongoing struggle to imagine and institutionalize a new energy infrastructure for south Florida is a case in point. Holding citizens' conferences on major technology issues (e.g., GMO policy), as is done in Denmark and elsewhere in the EU, is another.[45]

---

43. I have taken the phrase "dark green" from Allen, *Francis Miracle,* 158. As suggested in chap. 5 above, the creative-middle stance I've elaborated upon here reflects the position staked out by Francis in *Laudato.*

44. See O'Brien, *Making Better Environmental Decisions.*

45. For further examples, see Whiteside, *Precautionary Politics,* chap. 5.

An evolutionary ethics of responsible care and creative-middle approach provide ethical and conceptual resources for persons of faith and others involved in participatory technology assessment and other deliberative-democratic practices. In these contexts, such resources can help experts in many fields and laypersons from all walks of life find common ground as they deliberate over the purposes and potential trajectories of new technologies. During these deliberations, one must hope the prior and more compelling eco-social question concerning the good life and human/planetary flourishing will surface and be given serious consideration. For what is most needed is a more radical reform *of* technology, which we have said begins by recognizing our complicity with the debilitating pattern of technology and leads on to its principled restraint, i.e., to putting devices big and small in their place by reengaging with focal things and practices. As we learn to restrain technology's reach, we will create the right context and develop the right habits and frame of mind for pursuing reforms *within* technology.

Accordingly, a politics of engagement and time committed to a radical reform of technology will pursue two related tasks. One is intentional refiguring of the public sphere so that focal practices may flourish in a "republic of focal concerns" (Borgmann). Another is reforms within technology (and economy) that complement or support the principled restraint of technology. Both an evolutionary ethics of responsible care and creative-middle approach may be useful in shaping reforms within technology, provided a prior and more basic commitment has been made to the principled restraint of technology. A few clues as to how we might summon and sustain the energy and creativity to undertake reform initiatives are given in the final chapter.

10

Beyond the Device Paradigm: A Theo-Etiquette of Tango

> The changes to our lives will be ongoing and large and will require uncommon nimbleness, physically and psychologically.
>
> —BILL MCKIBBEN

> Quantifiable time is grasped always from the outside. It is the has-been of an event, measured in retrospect. Event-time, on the other hand, is the feltness of experience in the making. One is measured after the fact, one is felt in the now of movement-making . . .
>
> —ERIN MANNING

> I love the tango a lot. It is something that comes from inside me.
>
> —JORGE MARIO BERGOGLIO

Executives often speak of how challenging it is to navigate a volatile, uncertain, and complex business environment. At the ready, management gurus counsel a more collaborative, agile style of leadership. Market success calls for connecting and innovating, not command-and-control. The winning formula: attract talent, empower high-functioning teams, and swap smart machines for human labor wherever possible.

Environmentalists say a world of extreme weather, rising ocean waters, falling agricultural output, and swelling numbers of environmental refugees will require leaders and followers alike to become more adaptive and resilient, to pivot toward post-carbon technologies and adopt more sustainable lifestyles. McKibben is right: the changes ahead will be ongoing and large; skillfully navigating a risky, runaway world will require uncommon nimbleness and creative movement-making on the floor of history. Albert Borgmann and Pope Francis are right as well: we need to confront our complicity with the pattern of technology if we hope to pivot and pilot our way along a different path. This chapter opens, accordingly, with a critical reflection on technological culture as a preface to exploring in later sections what lies beyond the device paradigm.

Inside Glass Cockpits

Technology-centered automation not only renders many jobs redundant but also de-skills remaining workers, denying them a sense of autonomy and opportunities to develop skills and master challenges. Consider the fate of pilots. Air travel has never been safer, and we have computerization to thank for that. Yet commercial pilots, who touch the controls for only a few minutes at take-off and landing, now function as computer operators within high-tech glass cockpits bedecked with monitors. They struggle to gain flying experience and maintain the skill, expertise, quick reflexes, and habits of attentiveness required when, as sometimes happens, the auto-pilot glitches and the plane and its passengers are back in their hands. Sophisticated simulation training only goes so far. At least one veteran pilot has stated flatly, "We're forgetting how to fly."[1]

According to Nicholas Carr, the shadow side of auto-pilot is emblematic of a larger, worrisome trend as digitalization and automation take hold of the controls: "The mounting evidence of an erosion of skills, a dulling of perceptions, and a slowing of reactions should give us all pause. As we begin to live our lives inside glass cockpits, we seem fated to discover what pilots already know: a glass cockpit can also be a glass cage."[2] Carr's allusion to Weber lends weight to the increasingly common sentiment that we are fated to living in a world of smart machines all connected in the Cloud. Google and Big Auto, for example, aim to create a world of self-driving cars, and they say only fools and the Amish would reject the safety and freedom to do other things while mobile. Yet just when the uncommon nimbleness

1. Carr, *Glass Cage*, 15.
2. Ibid., 63.

and other capabilities required for a deft handling of the increasingly severe conditions we're fast flying toward are needed most, we are counseled to kick back for a Cadillac ride and let the learning algorithms take over.[3] How long will we be taken for a ride?

The brilliant coders, gifted engineers, and savvy entrepreneurs of Palo Alto and similar environs bringing to us the Segway Advanced Portable Robot, Modular TV, DietSensor SCiO Food Scanner, Parrot Disco Drone, Paro, and a thousand other wonders of the automated life are, by and large, a likeable and well-meaning lot. They work hard at their lovely jobs, often find themselves in a flow state, and sometimes create socially and therapeutically useful products. Yet elsewhere a good many folks remain stuck in lousy jobs and subject to nickel-and-diming by myopic management. A now run-of-the-mill alienation registers annually in high levels of employee disengagement. During his pontificate, John Paul II stated emphatically that such a situation is unacceptable, that work should count for more than a paycheck, yet millions are still counting the minutes till they punch the clock.

Off the clock, growing numbers find themselves estranged from rhythms and intimacies that fail to move with the speed, power, and dexterity of, say, a video game such as *Bloodborne*. Glass cockpits of a sort deliver them from the elements and flesh-and-blood entanglements into ersatz heroism, cyber-teamwork, and worlds of stunning verisimilitude more exciting and less burdensome than, say, a sneezy walk among the monuments and cherry blossoms. History and nature under glass, please.

What unites privileged and plebe these days is not shared experiences of military service, church, or school but love of video games. The mass exodus from the quotidian into virtual reality is accelerating, the demographic becoming more diverse. Active gamers now number nearly 200 million in the US, joined by 1.8 billion *virtualistas* worldwide; collectively, gamers spend more than three billion hours a week in their customized glass cockpits. Whence cometh the hunger behind the games? Whereas opportunities for intense engagement, grand purpose, and creative collaboration are lacking in the Muggle realm, they abound in *Call of Duty, Madden NFL 16, Fallout 4*, and the like. For many gamers, we are told, "the real world just doesn't offer up as easily the carefully designed pleasures, the thrilling challenges, and the powerful social bonding afforded by virtual environments."[4] Nudged along by ingenious game mechanics, one can play the wizard-hero without actually risking anything or helping anyone. What is denied at

---

3. On the quest among data scientists for the ultimate learning algorithm, see Domingos, *Master Algorithm*.

4. McGonigal, *Reality Is Broken*, 3.

work—autonomy, mastery, purpose—can be accessed virtually for days and nights on end. The new slogan: "Workers of the world, get a Second Life!"[5]

However we characterize these glass cockpits and the *modus operandi* of those ensconced within them, two things stand out. One is the distancing from nature, bare feet no longer touching blessed earth, attentive to the aliveness of matter. Inside a glass cockpit one is weightless, disembodied, ungrounded; bodily presence recedes to a vanishing point as enthrallment takes hold. Levels of awareness and concern about the dangers involved ranges from the acute to the clueless, but the cultural tendency clearly moves in the direction of a blithe technophilia, with a corresponding atrophication of biophilia.

Yet with shallow experiential roots in earth's rich soils, we are less likely to know and care about what happens "out there" on our behalf; instead of wonder and gratitude, complacency and entitlement set in. Nor can we know our *place*, in the double sense of sensitively inhabiting particular landscapes and, philosophically, grasping more profoundly our place and role in the larger scheme of things. In *The Spell of the Sensuous*, magician and philosopher David Abram reminds us that embodiment, and embodied cognition, are fundamental to human flourishing:

---

5. Can deliverance from glass cockpits come through games that move beyond escapism and help us to develop the habits and skills needed to strengthen family and community life, and even tackle larger social problems? In *Reality Is Broken*, game designer Jane McGonigal does not run from the world-house on fire or ignore how 24/7 post-industrial speed-up leaves growing numbers starved for joy. And McGonigal's sense of the adjacent possible is nothing if not ambitious. "Imagine a near future," she says, "in which more of the real world works like a game," i.e., where the energy, focus, and skill currently devoted to video games is redirected to improving society. To her credit, McGonigal recognizes the power of gameplay to spark creativity and cultivate various human capabilities. Yet her concrete-utopian wager that a gamer generation can develop leadership abilities for managing the planet with greater skill and foresight by playing "alternate-reality games" such as *Spore*, *World Without Oil*, *Superstruct*, and her own *EVOKE*, and then apply these capabilities (along with the game-design principles they've learned) to complex social realities, is problematic. The bugs in this software aren't just entrenched power structures deeply resistant to democratic-egalitarian alternatives hatched by subversive game developers, or the inconvenient fact that the most sophisticated, popular games on the market are under corporate control and hence carry a mandate to maximize profits, not start a revolution. Davids, after all, do take down Goliaths on occasion. No, the other thing bugging me is a suspicion that McGonigal's hip alternative, for all its creativity and sincerity, is itself only a design engineering framework for a new version, a new release, of the device paradigm. What we are likely to get, I fear, is another kind of glass cockpit rather than a more radical reform and principled restraint *of* technology. In short, gamifying life and work along "progressive" lines will not liberate us from the glass cage.

> Without the oxygenating breath of the forests, without the clutch of gravity and the tumbled magic of river rapids, we have no distance from our technologies, no way of assessing their limitations, no way to keep ourselves from turning into them. We need to know the textures, the rhythms and tastes of the bodily world, and to distinguish readily between such tastes and those of our own invention. Direct sensuous reality, in all its more-than-human mystery, remains the sole solid touchstone for an experiential world now inundated with electronically-generated vistas and engineered pleasures; only in regular contact with the tangible ground and sky can we learn how to orient and to navigate in the multiple dimensions that now claim us.[6]

An evolutionary ethics of responsible care is renewed and strengthened through focal practices such as gardening and hiking that bring us into direct, sensuous contact with the blooming, buzzing ecologies we inhabit. Wonder, gratitude, humility, joy, generosity, care—all are evoked by encounters with a more-than-human world, whether they occur in the backyard or in wild places of commanding presence such as Yellowstone. Entry into the Anthropocene need not entail disenchantment with a no-longer-pristine nature or make reenchantment dependent on game designs and wonderful gadgets.[7]

A closely related and equally disquieting characteristic of life in glass cockpits is the lack of vital movement, dialogue, and intimacy. For all the action on the screen and with the thumbs, there seems to be no room or desire for, say, a little cheek-to-cheek dancing. Hence we need not attend to, much less be accountable to, the *actual* other seeking eye contact, a kind

---

6. Abram, *The Spell of the Sensuous*, x. Abram's plea for reengagement with nature is yet another example of what Doran calls resistance to the "mechanomorphic distortion" of the dialectics of community and culture.

7. Cf. Bennett, *Enchantment of Modern Life*. Bennett expresses wariness toward religious and theological traditions that, in her view, falsely encompass the experience of enchantment and wonder within a teleological vision or make it dependent on an encounter with the supernatural. Still, her onto-story and "enchanted materialism" shares an affinity with the process-oriented, eco-theological perspective informing these pages. I share her conviction that "strong" versions of the modern disenchantment narrative produce unwelcome effects and that our capacity for joy, which expresses a "profound and . . . primordial attachment to being" (169) and is experienced as gift rather than achievement, is a vital source of ethical engagement and generosity. And I affirm her intuition that many sites of joy still exist in the modern world; states of beatitude may blossom forth in the strangest places, as Rumi and other mystical poets remind us. However, I am less sanguine than her regarding the ethical-practical effects of enchantment—and enthrallment—with computers, robots, and other technological devices.

look, a helping hand, or the warmth of a close embrace. Such physical and emotional distancing allows us to deny that we all "remain ultimately answerable to the concrete ethical exigency of the *face to face relation*."[8]

Perhaps it is all too risky. Today's customized self-enclosures seem designed as much for protection or simply release from the demands of fleshy-needy others as for anything else. In a 2015 Pew survey, "nearly half of eighteen-to-twenty-nine-year-olds said they used their phones to 'avoid others around you.'"[9] Little wonder that some take entrepreneur and computer scientist David Levy seriously and look forward to the day, allegedly not far away, when "love with robots will be as normal as love with other humans, while the number of sexual acts and lovemaking positions commonly practiced between humans will be extended, as robots will teach more than is in all the world's published sex manuals combined."[10] Note here the emphasis on the numbers and who the partners and teachers are. In reply to the likes of Levy, psychologist Sherry Turkle suggests a culture so infatuated with technology and so unenthusiastic about enduring attachments may be suffering from "an emotional dumbing down, a willful turning away from the complexities of human partnerships—the inauthentic as a new aesthetic."[11] A growing preference for and preoccupation with continuous, quick-and-easy exchanges conducted through social media ("gotta run, just shoot me a text") may have a subtly corrosive effect on relationships, if only because it renders us less *present*—i.e., more distracted and less available—to others in the day-to-day. A social-media bubble is a glass cockpit by another name. As photographer Eric Pickersgill's brilliant exhibit "Removed" unveils, we're becoming people of the device who are "there but not there."[12]

With so much leisure and love commodified, and with more work automated or dumbed down, many of us keep intimacy at arm's length and lose touch with more engaging focal things and practices, alienated from our most distinctive capabilities and deepest desires. The confusion around companionship and sex with sociable robots is a distressing sign of the times as is the mass retreat into virtual game worlds. Where the central vacuity of modern technology's prevailing pattern either goes unnoticed or is denied, we drift away from each other and find it more difficult to discover what is truly fulfilling.

8. Kearney, *Wake of Imagination*, 365.
9. Weisberg, "Hopelessly Hooked."
10. David Levy, quoted in Turkle, *Alone Together*, 5.
11. Ibid., 6.
12. Ibid., 14.

Viewed sociologically, these forms of alienation, confusion, and debility may be read in part as the negative byproduct of modernity's self-propelling process of social acceleration, a process that yields the paradox of a "frenetic standstill" (Rosa). As Turkle notes, it's a self-reinforcing system: "If the problem is that too much technology has made us busy and anxious, the solution will be another technology that will organize, amuse, and relax us."[13] The treadmill of production and consumption conditions us to embrace the next release, the next big thing, all the while distracting us from opportunities to develop character and talent, connect meaningfully with others, and contribute to the common good through participation in communities of focal practice that sustain a living tradition and generate the renewable, psychic-spiritual energy that makes a sustainably abundant life possible.

Let's pivot by posing a practical question: Absent a handy eject button, how are we to exit these glass cockpits? Put another way, how do we disengage from the device paradigm? In earlier chapters I broached these questions through discussions of focal practices, meaningful work, fruitful leisure, and the priority of the cultural. In what follows I offer a descriptive account of one focal practice—Argentine tango dancing—from the perspective of a participant-insider, i.e., a *tanguero*.[14] Viewed as a social dance (distinct from a performance art with professional stage shows, elite competitions, etc.), tango offers abundant insight into how we might learn to dance well with each other—and with technology—in communities attuned to the dialectic of tradition and innovation. Tango's dynamic synchrony, close partnerships, and intrinsic temporalities tutor us in resistance to the device paradigm, enabling us to move beyond it by connecting creatively and intimately in body, mind, and spirit with others.

I first survey illuminating moments in the social history of tango, a history with valuable lessons for practitioners of a politics of engagement and time today. I then invite readers to take in a *milonga* (tango dance party) from the vantage point of a dance-hall balcony in Buenos Aires, not simply to survey the contemporary tango scene but to step into the shoes of *milongueras y milongueros* so as to feel the floor—and the event-time of the dance—as they do. I close with brief reflections on what I call the

---

13. Ibid., 11.

14. The rest of this chapter may be read as an exercise in what Borgmann calls "deictic discourse," a language of testimony meant to disclose a focal concern. Often taking poetic or literary form, deictic discourse illuminates "what concerns me and, if successful, provides you with an understanding that [may] move you to act as I have been moved" (*Power Failure*, 180–81). Or in this case, move you to dance.

"theo-etiquette of tango" and what it takes to navigate the floor of history with grace and power.

Blues of Buenos Aires

Tango's sociocultural origins can be traced to working-class tenement life in Buenos Aires during the late nineteenth century.[15] The *barrios* and sizable immigrant populations inhabiting them were products of an international division of labor. From 1860 onward, "the pull of the European economy transformed what had been a backwater of the Spanish Empire into the most successful export economy in Latin America."[16] As the country's immense natural resources became more accessible through an expanding railroad system, Argentina's economic elite lacked one thing: cheap labor. Active recruitment of migrant workers ensued. Lured by promises of land and work and supplied upon arrival with a week's worth of food, young men from Naples and other southern European provinces arrived in Buenos Aires to seek their fortune and, hopefully, return triumphant to the mother country after a few years. The city absorbed three million immigrants between 1871 and 1914. Most never returned, and most never ascended the Argentine social ladder, the upper rungs of which became more exclusive over time.

In the *arrabales* (slums at the city outskirts), males far outnumbered females. Only live music existed and, for the poor, was to be found only in three places: brothels, street corners, and tenement courtyards. Gathering around the music, day laborers, waiters, sheep-shearers, tradesmen, sailors, ex-gauchos, *compadritos* (hoodlums), and other working-class men from different cultural backgrounds invented tango, learning from and dancing mostly with each other in the hope of winning a bride impressed by a breadwinner with game. In these circumstances, where eligible women were

---

15. Tango has remote roots in African dance traditions, and the black creole element has continued to shape its evolution; see Thompson, *Tango*. According to Thompson, the early tango culture of Buenos Aires "emerged from the encounter of dance concepts from Kongo with the city's cultural and social situation, involving African-born blacks, blacks born in Argentina, European migrants from Spain and Italy, and Europeans born in Argentina, including ex-cattlemen, some of them black, in from the pampas, looking for work" (8). In *Muntu in Crisis*, Boulaga says that "periodicity is the substantial time of things . . . Everything is alternation and rhythm . . . Rhythm is vital . . . It is rhythm which produces ecstasy, that flowing out of one's self that is identical with the vital force . . . It would not be exaggerated to affirm that rhythm is the architectural framework of the self, which for the [African] human being . . . is the most fundamental experience, which eludes all of the trappings of malign genius, which remains free of all doubt, and which is *Je danse, donc je vie* [I dance, therefore I am alive]" (50).

16. Hora, *Landowners of Argentine Pampas*, 1.

few and one's life chances limited, working-class *porteños* (Buenos Aires residents) acted creatively to keep the wolves of loneliness and frustration at bay, and to improve their odds of finding love. They answered the social question by converting "the outrage of the years into a music."[17]

At its simplest, tango was about the long shot of getting the girl. Yet there is more to the story than male desire, bravura, bravado, and heartbreak. Within the *conventillos* (tenement blocks), tango was the "music of the poor" bringing men and women, young and old, together to dance and enjoy themselves, to experience moments of joy in trying circumstances. Not far removed from their European provincial roots, immigrant *porteños* adapted to urban overcrowding by reinventing tradition, with Afro-Argentines strongly influencing the creative process despite their small numbers. Early tango musicians experimented with and blended rhythms and melodies from candombe, vals criollo, habanera, flamenco, polka, milonga, mazurka, and contradanse. During the early 1900s guitars, violins, flutes, and clarinets were joined by the tango's signature instrument—the bandoneon—a concertina that exudes a sensual and tragic sound. Trios began playing in the growing number of dance academies and cabarets opening up. Tango dance innovators adapted the couple dance hold used in waltz and polka (both of which were considered "new and daring" dance forms at the time) and added a revolutionary element: improvisation. What has been dubbed the "blues of Buenos Aires" was, in its formative decades, a vibrant expression of working-class cultural agency, artistic creativity, and multicultural community formation under conditions of poverty, discrimination, and political repression.[18]

Tango-Mania

Prior to World War I, the tango lyrics of music-hall and nightclub performers spoke of knife fights and lost love, yet typically they were lighthearted, comedic, and often bawdy, reflecting a street-savvy sensibility and sometimes satirical outlook on modern life. Tango lyricist Angel Villoldo's "The Bicycle" (1910) poked fun at those bedazzled by the onrush of new inventions:

> In this day and age
> We have this rage
> For everything electric

---

17. Jorge Luis Borges, quoted in Thompson, *Tango*, 3.
18. Castro, *Argentine Tango as Social History*.

> The microphone,
> The telephone
> The panpirilintíntophone
> Plus cinematography
> Biography, caustography,
> Pajalacafluchinography,
> Not to mention chingatapuchinography.[19]

As historian Robert Farris Thompson notes, Villoldo's "electro-obscenities" humorously put technology in its place, playfully mocking "the pretensions of modernism."[20] Tango's jousting, playful spirit in these days is captured in a photograph from 1912 showing strikers, rugged men in suspendered pants rolled up, dancing away with each other in ankle-deep beach water: bread, roses, and tango.

Then the sons of the Argentine nouveaux riche let the cat out of the bag, introducing to Parisian society the tango they'd learned (likely while slumming). In 1912, tango took Paris salons and soon after the world by storm. From New York to Shanghai and every big city in between, "tango-mania" broke out. Chattering-class hunger for the exotic played a role, yet the new dance was undeniably magnetic. While the Kaiser ordered court-marshals for any officer caught dancing tango in uniform, the Czar hired Argentines to instruct the royal family. When advisors to Pius X urged a ban, the pope asked to see the dance. The demonstration couple kept it on the down-low, prompting Pius to quip the folk dances of his home village were more fun. (Not to be denied, conservative zealots at the Vatican penned a pastoral letter in 1914 on behalf of His Holiness, which reads in part: "The tango, which has already been condemned by illustrious Bishops and is prohibited even in Protestant countries, must be absolutely prohibited in the see of the Roman Pontiff, the centre of the Catholic religion.")

Although tango-mania horrified elite *porteños* back home, they subsequently felt obliged to co-opt and domesticate the "dance of the brothels" once the Great War ended. A now fashionable tango was reimported, and in the postwar years tango lyrics were sentimentalized as Argentina's middle and upper classes caught on and created a market for a smoother tango sound. Carlos Gardel and other great tango singers took center stage. From the mid-1920s to the mid-1930s, tango orchestras and marquee singers entertained a growing audience in live performances, on the radio, and in film.

---

19. Angel Villoldo, quoted in Thompson, *Tango*, 28–29.
20. Ibid., 29.

During these years, the music became more complex and tango dancing languished.

Golden and Dark Ages

It's an early evening in 1935 and no customers have arrived at the club. Both economy and politics have turned ugly during what comes to be known as Argentina's Infamous Decade. The orchestra takes advantage of the empty room and band leader's absence to jam a little, much to the delight of the staff. Rodolfo Biagi, the pianist, has infused energy into the band's sound with a crisp, clean, foot-tapping rhythm—like nothing that's been heard in Buenos Aires for over a decade.[21] After conductor Juan D'Arienzo appears and folks roll in, the staff wants to hear the "new tune" again. D'Arienzo is strict about arrangements, and the band members are nervous about straying from what the master expects. But when Biagi kicks in and the crowd loves it, the conductor quickly approves. D'Arienzo's orchestra lands a recording contract not long after, and their old-new dance beat proves a big hit. Soon the dance salons and social clubs are humming again.

Most tango historians view this moment as the beginning of tango's golden age, a twenty-year period in which a synergy of music, song, and dance met with wide appeal—everyone was now dancing—and resulted in an exceptionally high level of artistry. Popular orchestra leaders such as Aníbal Troilo found a way to include gifted singers without showcasing them and taking away from dance rhythm. Dance styles ranged from elegant and restrained in the city's tonier northern neighborhoods to earthier and "provocative" in rougher southern sections. Carlos Alberto Estévez (nicknamed "Petróleo") and other Club Nelson Men experimented with new figures, reintroducing a more playful element into the dance.

At mid-century, men still outnumbered women at the *milongas*. By convention, the place to meet a girl and win her over was on the dance floor. The message to young men was clear: "It is just as important to be well danced as to be well versed or well read."[22] Regardless of where one danced or what style one preferred, mastery of technique was paramount and took sustained effort. Hence the all-male *práctica* tradition was revived (women had their own training sessions, mostly in the home with fathers and mothers teaching their daughters). Beginners went to local clubs for instruction several times a week, first learning the "woman's part" through a rigorous apprenticeship to older dancers typically lasting nine months before train-

---

21. Denniston, *Meaning of Tango*, 68.
22. Thompson, *Tango*, 5.

ing as a leader began. At the *prácticas*, accomplished *tangueros* tutored novices and continued to refine their own skills by partnering with other men. The best male partners gave demonstrations. A novice's training took no less than three years before permission was granted to attend a *milonga*. This was the cultural milieu for Jorge Mario Bergoglio (b. 1936), who grew up during tango's golden age and enjoyed many a dance with friends before joining the Jesuits.

Juan Perón's populist-authoritarian government (1946–1955) encouraged the tango dance craze as an emblem of national identity and pride. Entire soccer fields were filled with dancers. While Perón's repression of political opponents (left and right) claimed a few leading tango figures—notably the pianist Osvaldo Pugliese and Eva's film rival, Lamarque Libertad—many tango singers and performers were strong supporters of a Peronist agenda that led to social-democratic reforms, strong labor union growth, and workers' increased share of national income during this time.

Then in 1955, when Perón was ousted in a military coup, a nearly thirty-year tango "dark age" began. All things Peronist came under suspicion, including the "filthy" tango. Curfews were imposed, shutting down club and dance venues. A succession of military juntas and unstable governments welcomed the arrival of other Latin beats as well as rock n' roll music. Tango went out of fashion among young Argentines, who flocked to the new rock concert events and left the *prácticas* behind. The only innovation was avant-garde tango music led by the brilliant Astor Piazzolla, loved more by Parisians than *porteños* in the postwar decades. As more tango salons, cafés, and social clubs closed, the dancing went underground. Many of the best *tangueros y tangueras*, carriers of the living tradition, grew older without passing their passion on to the next generation. Especially during the generals' "dirty war" against dissidents (1976–83), the country was frozen culturally and social life was paralyzed by fear.

Open Floor

With the end of military rule in 1983, Argentina entered a period of more stable democratic life. The new government set up neighborhood cultural centers to revive civic activity. Tango floors opened again and a new generation began to seek out the old masters. Nor was the tango renaissance limited to Buenos Aires. The year 1983 also marked the opening of the highly successful show *Tango Argentino*, which toured the world leaving fans behind in each city eager to learn the dance. For years the level of dancing remained quite low compared to golden-age standards; the intensive apprenticeship model of the *prácticas* yielded to group lessons, often taught by teachers with relatively little tango training and experience. Still, by the new millennium a world tango scene had emerged with Buenos Aires at its center, new blockbuster shows such as Luis Bravo's *Forever Tango*, thriving dance communities in dozens of cities, and a new wave of tango dance masters, some of them *tango nuevo* innovators.

Seeking to extend the range of physical movement possible in the dance, *tango nuevo* artists experimented with exceptions to and variations on the traditional tango dance vocabulary. During the 1990s, Gustavo Naveira and others developed a new understanding of the dance's physical mechanics and a new teaching method featuring an eight-count basic figure more accessible to large numbers of beginners arriving on the scene. Instead of teaching set figures (or "words"), the new method focused on a simple "alphabet" of forward, side, and back steps. This shift opened up a wider range of step combinations—new figures and patterns—and thus renewed tango's spirit of improvisation. The experimental focus and sense of new possibilities led to a revival of the *práctica* tradition.[23]

Musical horizons have expanded as well. Gotan Project, Otros Aires, Tanghetto, and other neo-tango musical groups blend electronic and acoustic sounds. On "alternative" floors, DJs play blues, R&B, and other music amenable to *tango nuevo* dancing. In turn, these new developments have sparked a strong "traditionalist" revival with renewed fidelity to the milonga and tango salon styles perfected during the golden age as well as attempts to recover older forms such as tango criollo and canyengue.

What Smiles Say

Now imagine yourself at a dance hall in Buenos Aires today. It's just after midnight when the better dancers are starting to appear on the diamond-patterned floor. From a balcony, you survey a dozen or so couples dancing counterclockwise around the perimeter to a tango Vals (tango Waltz), a piece called "Pobre Flor" ("Poor Flower") composed by Alfredo de Angelis in 1946. The multi-sensory gestalt is lovely: dancers form a smooth-flowing river of movement. It's the last dance of a *tanda* (set of three to four tango songs), to be followed by a short break called a *cortina* when refreshment and re-pairing occurs.

Turning your attention to individual couples, you first notice the energy radiating from one pair. These dancers have discovered what moves

---

23. Merritt, *Tango Nuevo*. Master instructor and world-class dancer Andrés Amarilla has systematized the combinatory possibilities of tango dance steps in a code.

them, each step coming from the inside out, the two creating a magnetic field. There's a refined artistry on display as well, an array of figures and turns in three time, with crisp cuts one way and then another. Dancing fearlessly, they move across the floor like they own it, yet without showiness. No wonder the follower wears a confident smile. Meanwhile, an older couple takes simpler and more measured steps—nothing fancy here. While they seem to ripple quietly across the floor, you sense a deep undercurrent of feeling. The follower smiles serenely, relishing the close embrace and clarity offered by a quiet, confident lead. Two hearts synchronized are worth more than a thousand figures, and they both know it. A third pair seems entranced within a circle of their own making, exuding a lush elegance with every exquisite step. Before long you are lost in their liquid motion. How can they move so smoothly and so closely in unison? Somehow they've melded with the Vals music. And here too a serene smile seems as fitting as the last. The fourth pair you spy seems different from the others. Their longer steps, more open embrace, and playfulness mark them as *tango nuevo* dancers. Like a gyroscope, they seem to generate graceful twists and turns effortlessly while staying on axis. And when the lead does take the follower off-axis in a *volcada*, the weight distribution remains perfectly balanced. You sense a lighter, more buoyant movement unfolding.

Still another couple impresses you with how effortlessly they move together, imperceptibly exchanging the lead at different moments, as if in intimate conversation with one another. And there's stillness in how they step, a meditation in movement. Both smile serenely, and no one here blinks at two women dancing together with their eyes closed. Finally, your eye turns to royalty. Well, she's certainly being treated that way. At one point the leader sweeps a free foot around to his left, his toe drawing a *rulo* (circle of light) on the floor while he simultaneously invites the follower into it. And she accepts, embellishing her step just so along the way. She shines a moment in the circle of light and then exits as gracefully as she entered. Through every phrase their steps are smooth and luxurious. ("He's a Cadillac," the women say.) Her smile says it all.

And so the *tanda* ends. Both dancers smile, thank each other, and part ways. During the *cortina*, there's a minute to reflect on what it takes to co-create such beauty in motion.

Good posture and frame, command of technique, a fluid knowledge of tango's syntax of movement, deft navigation of a crowded floor—these competencies and more distinguish advanced from beginner and intermediate leads. Yet ask followers what they want most of all, and the unanimous reply is a good connection and staying with the beat. The intimacy of the basic dance position—two people standing directly in front of one another, shoulders parallel, bodies aligned, ready to embrace—reflects what is most important: two hearts coming close. Connection and musicality are what make for a joyful experience, whatever the level of competence.

You can always tell a good lead by the follower's smile. She feels safe, cared for, and thus moves confidently and with ease. Leaders must extend trust and earn it. Attentiveness to the music and solicitude toward one's partner is paramount. Force the step, and the connection will fade or never be made. The follower stiffens, blocks you out. But lose yourself and you find a way together across the floor. Tango teachers say the secret to leading well is in following the follower.

That said, the two cannot dance as one without each one maintaining good posture, a firm yet flexible frame, solid technique, and a certain comportment. Posture and frame are enabling structures crucial to dancing well, and the same must be said of technique. The cat-like walking step alone requires months of practice before one approaches proficiency; dancers spend a lifetime mastering it. As Erin Manning reminds us, "the necessity of technique should not be underestimated: technique is what allows the ecology of movement to open itself to its generative potential."[24] No technique, no skillful improvisation and creativity. With mastery of technique comes freedom of expression. Yet equally crucial is comportment, i.e., certain refined habits that convey character. It is first and foremost through their comportment that tango dancers announce and make felt their convivial presence on the floor. In tango, ethics begins with etiquette, which in turn complements style.

Dancers develop skills and habits through disciplined education, a social learning structured by master-apprentice relationships. Recall the golden-age *prácticas*. Mastery requires dedication and comes slowly through countless hours of practice. Role modeling and instructive feedback from one's teachers is indispensable (but so is a mirror and self-critique). All this makes possible the flow of energy between partners as the dance unfolds. Within a solid and sensible frame, and with mutual sensitivity and respect, steps and heart beats synchronize in the event-time of movement-making.

Improv performers often begin their training with the well-known "Yes, and . . ." exercise. I serve up a declarative statement and you return with a "Yes, and . . ." reply that picks up on a word or phrase I used and imaginatively strokes it somewhere else. Then with my own "Yes, and . . ." volley I stroke one of your words or phrases somewhere else, the verbal back-and-forth bouncing along wittily and often into ludicrous, fantastic spaces. Co-creativity made possible by an enabling structure. Imagination

---

24. Manning, *Always More Than One*, 78.

runs wild because it follows the rules of the game. A verbal entanglement births a smile and a laugh. And in each moment, the adjacent possible draws the pair forward.

As with Improv, tango social dancing is never choreographed. Within the enabling structures of the dance, leader and follower interpret the music variously, some couples tuning into a single instrument's sound, others to the often elusive rhythm (traditional tango orchestras have no percussion). And with each step, leader and follower convivially co-create a phrase in response to the music and to each other. Unlike Improv, though, where jarring juxtapositions continually appear to our surprise and delight, the *novum* of each tango dance step unfolds within the texture of the music and in a manner more akin to a rich, intimate conversation between two friends. Its affective range includes the flirtatious and playful yet far exceeds it.

Tango's reputation as the language of love *par excellence* is well-deserved. To dub the dance "passionate" or "melancholic" and move on conceals a great deal, though, for the layers of feeling and desire expressed and evoked during the dance range so widely—from the elemental and erotic to the exultant and ecstatic, from the blue to the buoyant. Much happens when two souls entwine and wander across the wood, lured on by the music. What plays out in this deep play cannot be prefigured. The two venture forth, stepping together again and again into new territory, exploring the adjacent possible through a movement-making at once composed and untamed, at times intensely radiant and in others more restrained yet no less intense.

Each dance thus is never the same, even when the steps and one's partner are familiar. After a good dance, as after a good conversation or a memorable walk down an unfamiliar path, one is never quite the same. In this way, tango invites us to become the person we had not a clue we could be, to become otherwise.

And what shall we say of the misstep or stumble, the awkward moment? An unclear signal from the lead, a rush into the next step by a nervous and inexperienced follower, poor technique—these common failures among the tango faithful turn a smooth synchrony into an unwelcome heaviness and tension. Tango etiquette is clear about these inevitable falls from grace: the lead takes responsibility—always—and repairs the situation immediately by adjusting as necessary and leading the pair back into lightness and ease. What makes a quick recovery challenging is the uncommon nimbleness of mind required: the lead has to reframe the situation on the fly and act with precision and timeliness, lest the dance drag or come to a crashing halt. There is no auto-pilot or game mechanics in tango, only grace under pressure.

At their best, tango communities across the globe practice a generous hospitality, embody the spirit of conviviality, and are comfortable with diversity. In a youth-obsessed culture, they are one of the few places outside of family life where different generations socialize and interact with each other in an atmosphere of mutual respect. Anthropologist and tango dancer Carolyn Merritt recounts the representative testimony of a novice *tanguera*:

> Speaking of one of the first *milongas* she'd ever attended, on the outskirts of Paris, she describes a place where people are truly seeing, listening, and understanding one another—free of the prejudices and judgments humans are so prone to; a place where superficial differences and categories like race and gender seem to dissolve. Transcending the bounds of the couple, what she describes is a communion that embraces the very place itself, and although only momentary, makes it one of limitless possibility.[25]

Merritt's anecdote may be multiplied many times over, and here it seems fitting to recall my own initiation into the mysteries of tango.

Beginner leads speak of "tango angels" appearing just when the difficulties of learning the dance seem insurmountable. At a Friday-night *milonga* in west Philadelphia, with only a few months of lessons behind me, I had the temerity to ask Lori, an advanced *tanguera*, to dance. Knowing she had taken on a work of mercy, Lori smiled and took my hand. Sensing the terror to which I quickly succumbed once the music began, she taught me to breathe when I led a step, to breathe into each step without fear, and to savor the connection with one's partner above all. Imagine that with each step, she said, you are entering a circle of light. Though Lori was following as I fumbled us ahead, in fact she was showing me the way to navigate a crowded floor with grace. To this day her imprint has remained, and I am forever grateful.

Years later Lori founded Sangha Space, a community organization dedicated to "connecting people through movement." Sangha Space hosts a weekly *práctica* and a beautiful *milonga* each month. Under an arched ceiling and often to live music, people of all ages and from all walks of life pair up and join a river of movement. It matters little that the fireplace at the far end of the dance hall is never lit, for the energy flowing throughout more than suffices.

Places like Sangha Space are training grounds for the transformation of selves, an *ecclesia* by another name.[26]

---

25. Merritt, *Tango Nuevo*, 112.
26. Commenting on Ivone Gebara's song of the earth, Keller imagines a milongic

River of Movement

The *cortina* is ending, what now? Every *tanda* begins off the floor in the moment when leader and follower agree to dance together. Traditional tango etiquette offers a subtle, face-saving way for leaders to invite followers known as the *cabeceo* (a nod). As the music for a new *tanda* begins, from a distance a leader will signal to a follower his desire to dance by catching her eye and nodding ever so slightly toward the floor. If the follower smiles in agreement, she often will wait while he crosses the room, takes her hand, and escorts her to the floor (the traditional "entryway" is a corner). And this is only the first giving and receiving of an invitation to dance: every subtle signal the lead will send, every little lure, is but an invitation graciously accepted, the follower's step itself a signal and invitation for a responsive lead, a moved mover. In freedom the two give and receive, artfully conveying intention and desire.

The floor begins to fill, a river of movement rises. This *tanda* is classic tango salon featuring the songs of golden-age great Carlos di Sarli. Dancing to "La Capilla Blanca" ("The White Chapel"), a different kind of elegance is on display. More elaborate figures appear, and while the flow is still counterclockwise around the floor as always, couples modulate the tempo. In some moments they dance slowly and in others come almost to a stop, hovering for a second before taking the next step. This focal moment, known as *la pausa*, is considered by some masters the essence of tango. In the gathering of *la pausa*, partners become more aware both of each other and of themselves. They dwell a little more deeply in the connection—a savoring of event-time, Manning might say. Yet simultaneously this "on hold" moment, this truly-being-held moment, is pregnant with possibility. During *la pausa* the partners also anticipate the next step. In this way, *la pausa* intensifies and crystallizes the felt experience of movement-making. The event-time of *la pausa* holds within itself the adjacent possible of co-creative movement.

Shifting back to a wider frame of vision, what strikes you again is the river of movement. Each pair expresses the music uniquely and creates an energy field all its own, yet at the same time the steps each couple takes are synchronized with all the other couples flowing counterclockwise around the floor. Such rivers possess power, moving those immersed in them into a self-forgetful state and the joy of communion.

---

refiguring of Paul's image for the church as the body of Christ: "Singing the song born of the body, arms and embraces, we become in truth 'members one of another' . . . we become part of each other in the interdependence of the body, a body 'called out,' ek-*klesia*, from the status quo of numbed disembodiment and bonds that block" (*On the Mystery*, 120).

Good *milongas* are convivial, co-creative realizations of immense beauty, celebrations of abundant life. Richly flavored repasts of shared movement and sensual-spiritual connection, they prefigure a Great Feast. One leaves such events with more energy, not less, and their recurrence over time within a vibrant dance community—one in which joy and gratitude consistently attend creative expression—is a notable instance of what I mean by sustainable abundance for all.

Can we imagine Sunday worship or the family meal in analogous fashion? If so, then we may be ready to move. Stepping onto the floor of history, we may be ready to join the river of movement that works toward making sustainable abundance for all a reality. Despite the burdens we will love the labor, I suspect, for it will be something that comes from inside us, something that lightens our spirit at each step.

In This Day and Age

The river of movement we need for human and planetary flourishing will not flow out of Silicon Valley, I wager, but rather from (re)engagement with focal things and practices we currently stumble over as we rush about. Were Villoldo writing "The Bicycle" today, perhaps he'd change the title to "The Self-Driving Car" and pen something along these lines:

> In this day and age
> We have this rage
> For everything digital,
> Robotic,
> And nano-secondal,
> The googling
> The texting
> The hyper-sexting
> Plus electron microscopy,
> Biopsy, warposcopy,
> Surveillobotomoscopy,
> Not to mention porntruffbuttoscopy.

Satire aside, it is worth recalling that a bicycle is a focal thing, and both casual bike-riding and the sport of cycling are focal practices. And the instruments a tango sextet plays are focal things, as are the dance hall and diamond-patterned floor upon which a river of movement rises, flooding conjoined hearts with joy.

By comparison, self-driving cars are—well, you get the point.

A Theo-etiquette of Tango

At present many of us grant the device paradigm too much sway and are the worse for it. Self-enclosed within glass cockpits, wholeness and flourishing elude us. Faced with this predicament, Borgmann suggests that

> a radical theology of technology must be alive to these debilities and follow them up to discover and determine their pattern, extent, depth, and consequences. What might emerge . . . is the insight that the technological style of life does have as an intrinsic feature the incapacity for salvation, i.e., for the wholeness of life. But [for the privileged] the decisive mark of human frailty is no longer, as in the pretechnological setting, manifest at the material level as hunger, nakedness, and sickness. Instead, it has become a crippling of our most profound capabilities and consequently a deprivation of things in their own right and depth. To be saved, accordingly, may involve the recovery of one's capacity for the fullness of nature, art, and for the pretechnological things and practices of daily life that lie half-buried under the surfeit of consumption.[27]

A good first step toward recovery is to admit we are complicit with the pattern of technology. Then perhaps isolation, social paralysis, and ennui might yield to co-creative movement as empathy awakens and imagination dares to envision the adjacent possible of sustainable abundance for all. Wherever we are, we can join the movement-making and learn how to lead.[28] Wherever we are, we can reaffirm our human vocation: the convivial co-creation of greater value and beauty amidst unfolding plenitude in universal history.

That vocation, we have said, takes concrete, dramatic form in and through the various roles we play and narrative identities we form within communities of focal practice. Such communities make possible recurring moments of artful performance characterized by wonder, joy, generosity,

---

27. Borgmann, *Power Failure*, 87.

28. Keller conveys the theological-ethical import—and dynamism—of a co-creative movement that reaches beyond any single community, any limited communion, to embrace with persistent love the related challenges of systemic change, sustainability, and social inclusion when she writes: "The co-creativity to which we are together lured produces more than togetherness: it effects the structures of justice that will support the creativity of an ever-diversifying togetherness. Our actions, our experiences, would be thus collected into a divine movement, an embrace that re/collects the whole variegated, struggling collective of creation" (*On the Mystery*, 125).

creativity, communion, and the expansion of human capabilities. Through disciplined engagement with and commitment to focal practices and the communities that sustain them, we become otherwise over time as we come to share more fully in the flow of human and divine love.

Where graciousness and gratitude abound, there the Spirit is movement-making. Adept tango partners and vibrant tango communities practice what may be called the *theo-etiquette of tango*: a gracious and joyful way of being in the world that combines embodied cognition, empathy, responsible care, imaginative daring, a desire to connect deeply, and creative fidelity to community and tradition. Always sensuous and sometimes sultry, tango dancing is above all a profound form of *deep play* and hence a site of the sacred. A sacramental quality infuses the movement, music, and close embraces felt during a good *milonga*. Through the artistry of an alluring lead one can glimpse God's gracious invitation to dance. Through the follower's exquisite reply one can witness a disciple's willingness to attend faithfully and in beautiful style. One can sense in the rivulets of joy flowing between partners the gathering of a great river, a joy unsurpassable. One can be led into a circle of light, led with care into a deeper and more intimate encounter with divine mystery.

Taking the Time

Asking how we spend our time, how we move through time, is another way of posing the question about the good life. In his sermons and writings, Pope Francis invites us to (re)learn the arts of pausing and reflecting, of minding the moment, of holding one another in close embrace. To create a truly humane, ecological culture requires the habit of "*taking time* to recover a serene harmony with creation, reflecting on our lifestyle and our ideals, and contemplating the Creator who lives among us and surrounds us, whose presence 'must not be contrived but found, uncovered'" (LS #225; my emphasis). By (re)learning the way of *la pausa contemplativa*, our spirits may "grow sweet again and fragrant and wild and fresh and thankful for any small event" (Rumi).

Given the cultural hegemony of the work ethic and the time-is-money logic of capitalism, it will take quite some time to realize the full potential of a post-jobs society in which we no longer spend our days rushing about breathless because we have discovered what activities are truly worth our time. These worthy activities—focal practices such as family dining, social dancing, common worship, and many more—put technology in its place. What is more, they are training grounds for teaching us how to be *leaders* the world truly needs. Our ability to move beyond the device paradigm into attunement with deeper rhythms, and to navigate the floor of history with grace and power, depends in no small measure on taking the time to engage in focal practices that *form* us and renew a radical generosity of spirit within us.

Good formation enables us to take up a moral politics of engagement and time, to pursue the adjacent possible of sustainable abundance for all. For in this moment near midnight the politics of fear and resentment must be met by a powerful countercurrent, a river of movement rising from deep within and rolling mightily across a fair land we may yet reclaim and renew.

Such rivers change the course of history.

Bibliography

Abram, David. *The Spell of the Sensuous*. New York: Vintage, 1996.
Acemoglo, Daron and James A. Robinson. *Why Nations Fail: The Origins of Power, Prosperity, and Poverty*. New York: Crown Business, 2012.
Ackerman, Diane. *Deep Play*. New York: Vintage, 2000.
Adam, Barbara. *Timescapes of Modernity: The Environment and Invisible Hazards*. New York: Routledge, 1998.
Allen, John L. *The Francis Miracle: Inside the Transformation of the Pope and the Church*. New York: Time, 2015.
Anderson, Ray C. *Confessions of a Radical Industrialist*. New York: St. Martin's, 2009.
Andolsen, Barbara Hilkert. *The New Job Contract: Economic Justice in an Age of Insecurity*. Eugene, OR: Wipf and Stock, 2009.
Angwin, Julia. *Dragnet Nation*. New York: Holt, 2014.
Arendt, Hannah. *The Human Condition*. Chicago: University of Chicago Press, 1958.
Ariely, Dan. *Predictably Irrational*. New York: Harper Perennial, 2010.
Aronowitz, Stanley, and Jonathan Cutler, eds., *Post-Work: The Wages of Cybernation*. New York: Routledge, 1998.
Arthur, W. Brian. "The Second Economy." *McKinsey Quarterly* (October-November 2011) 1–5.
Ashok, Vivekinan, et al. "Support for Redistribution in an Age of Rising Inequality: New Stylized Facts and Some Tentative Explanations." NBER Working Paper No. 21529 (September 2015).
Bales, Kevin. *Blood and Earth: Modern Slavery, Ecocide, and the Secret to Saving the World*. New York: Spiegel & Grau, 2016.
Barber, Benjamin R. *Strong Democracy: Participatory Politics for a New Age*. Berkeley: University of California Press, 1984.
Barks, Coleman, trans. *The Soul of Rumi*. New York: HarperCollins, 2001.
Baum, Gregory. *The Priority of Labor*. New York, Paulist, 1982.
Beck, Ulrich. *The Brave New World of Work*. Malden, MA: Polity, 2000.
———. *World at Risk*. Malden, MA: Polity, 2009.
———. *World Risk Society*. Malden, MA: Polity, 1999.
Beck, Ulrich, and Elisabeth Beck-Gernsheim. *Individualization*. London: SAGE, 2002.
Beigel, Gerard. *Faith and Social Justice in the Teaching of Pope John Paul II*. New York: Lang, 1997.
Benkler, Yochai. *The Wealth of Networks*. New Haven, CT: Yale University Press, 2006.
Bennett, Jane. *The Enchantment of Modern Life*. Princeton, NJ: Princeton University Press, 2001.

Bergmann, Frithjof, and Thomas Staehelin. *Starting With New Work: Creating a New Culture*. Online publisher: CreateSpace Independent Publishing Platform, 2014.
Berry, Thomas. *The Great Work: Our Way into the Future*. New York: Bell Tower, 1999.
———. *The Sacred Universe*. New York: Columbia University Press, 2009.
Biello, David. *The Unnatural World: The Race to Remake Civilization in Earth's Newest Age*. New York: Scribner, 2016.
Bix, Amy Sue. *Inventing Ourselves Out of Jobs? America's Debate Over Technological Unemployment 1929–1981*. Baltimore: Johns Hopkins University Press, 2000.
Bloch, Ernst. *The Principle of Hope*. Vols. 1–3. Cambridge, MA: MIT Press, 1986.
Borgmann, Albert. *Technology and the Character of Contemporary Life*. Chicago: University of Chicago Press, 1984.
———. *Crossing the Postmodern Divide*. Chicago: University of Chicago Press, 1992.
———. *Power Failure: Christianity in the Culture of Technology*. Grand Rapids: Brazos, 2003.
Bostrom, Nick, and Milan Cirkovic, eds., *Global Catastrophic Risks*. Oxford: Oxford University Press, 2011.
Botkin, Daniel B. *Discordant Harmonies: A New Ecology for the Twenty-First Century*. Oxford: Oxford University Press, 1990.
Boulaga, Fabien Eboussi. *Muntu in Crisis: African Authenticity and Philosophy*. Trenton, NJ: Africa World, 2014.
Bowles, Samuel, and Howard Gintis. *A Cooperative Species: Human Reciprocity and Its Evolution*. Princeton, NJ: Princeton University Press, 2011.
Boyle, James. *The Public Domain: Enclosing the Commons of the Mind*. New Haven, CT: Yale University Press, 2007.
Bradley, General Omar. "Armistice Day Speech (November 11, 1948)." In *Omar Bradley's Collected Writings*, Vol. 1 (U.S. Government Printing Office, 1967) 321–28.
Bregman, Rutger. *Utopia for Realists*. New York: Little, Brown, 2016.
Brockmann, Hilke, and Jan Delhey, eds. *Human Happiness and the Pursuit of Maximization: Is More Always Better?* New York: Springer, 2013.
Brunner, Grant. "Foxconn is attempting to replace its human workers with thousands of robots." *ExtremeTech*, July 8, 2014. http://www.extremetech.com/electronics/185960-foxconn-is-attempting-to-replace-its-human-workers-with-thousands-of-robots.
Brynjolfsson, Eric, and Andrew McAfee. *The Second Machine Age*. New York: Norton, 2014.
Brynjolfsson, Eric, et al. "Labor, Capital, and Ideas in the Power Law Economy." *Foreign Affairs* (2014) 44–52.
Callicott, J. Baird. *Beyond the Land Ethic*. Albany: SUNY Press, 1999.
Cannovo, Peter F. *The Working Landscape: Founding, Preservation, and the Politics of Place*. Cambridge, MA: MIT Press, 2007.
Carmin, JoAnn and Julian Agyeman, eds. *Environmental Inequalities beyond Borders: Local Perspectives on Global Injustices*. Cambridge, MA: MIT Press, 2011.
Carr, Nicholas. *The Glass Cage: Automation and Us*. New York: Norton, 2014.
Castro, Donald S. *The Argentine Tango As Social History, 1880–1955: The Soul of the People*. San Francisco: Mellen Research University Press, 1991.
Chambers, Iain. *Migrancy, Culture, Agency*. New York: Routledge, 1994.
Chenoweth, Erica, and Maria J Stephan. *Why Civil Resistance Works: The Strategic Logic of Nonviolent Conflict*. New York: Columbia University Press, 2011.

Christiansen, Drew, and Walter Grazer, eds. *And God Saw That It Was Good: Catholic Theology & the Environment*. Washington, DC: USCC, 1996.
Clark, Christopher. *The Sleepwalkers: How Europe Went to War in 1914*. London: Lane, 2013.
Clifton, Jim. *The Coming Jobs War*. New York: Gallup, 2011.
Cohen, Patricia. "Middle Class, but Feeling Economically Insecure." *New York Times*, April 10, 2015.
Connelly, William. *The Fragility of Things*. Durham, NC: Duke University Press, 2013.
Crompton, Tom and Tim Kasser. *Meeting Environmental Challenges: The Role of Human Identity*. Surrey: WWF-UK, 2009.
Daly, Herman E., and John C. Cobb. *For the Common Good: Redirecting the Economy toward Community, the Environment, and a Sustainable Future*. Boston: Beacon, 1994.
Deane-Drummond, Celia. "Joining the Dance: Catholic Social Teaching and Ecology." *New Blackfriars* 93 (1044) 155–62.
Denniston, Christine. *The Meaning of Tango*. London: Portico, 2007.
Despommier, Dickson. *The Vertical Farm: Feeding the World in the 21st Century*. New York: Picador, 2011.
de Waal, Frans. *The Age of Empathy*. New York: Broadway, 2010.
Doherty, Patrick. "How the Oil and Gas Industry Can Help Save the World." *Brink News*, October 17, 2016.
Domingos, Pedro. *The Master Algorithm*. New York: Basic, 2015.
Doran, Robert. *Theology and the Dialectics of History*. Toronto: University of Toronto Press, 1990.
Dorr, Donal. *Option for the Poor and for the Earth*. Maryknoll, NY: Orbis, 2012.
Dryzek, John S. *The Politics of the Earth: Environmental Discourses*, 3rd ed. Oxford: Oxford University Press, 2013.
Dussel, Enrique. *Ethics of Liberation*. Durham, NC: Duke University Press, 2013.
Edin, Kathryn J., and H. Luke Shaefer. *$2.00 a Day: Living on Almost Nothing in America*. New York: Mariner, 2016.
Ellison, Marvin M. *The Center Cannot Hold: The Search for a Global Economy of Justice*. Washington, DC: University Press of America, 1983.
Elsbernd, Mary. "Whatever Happened to Octogesima Adveniens?" *Theological Studies* 56 (1995) 39–60.
Eriksen, Thomas H. *Tyranny of the Moment: Fast and Slow in the Information Age*. London: Pluto, 2001.
Fallows, James. "The Planet-Saving, Capitalism-Subverting, Surprisingly Lucrative Investment Secrets of Al Gore." *Atlantic*, November 15, 2015, 23–29.
Feenberg, Andrew. "From Essentialism to Constructivism: Philosophy of Technology at the Crossroads." In *Technology and the Good Life?*, edited by Eric Higgs, et al., 294–315. Chicago: University of Chicago Press, 2000.
Fidler, Devin, and Marina Gorbis. "Technology: Prosperity by Design." *Democracy: A Journal of Ideas* (2016) 26–33.
Figueres, Christiana, et al. "Three years to safeguard our climate." *Nature* 546, no. 7660. June 28, 2017, 1047–53.
Finn, Daniel. *Just Trading: On the Ethics and Economics of International Trade*. Nashville: Abingdon, 1996.

Ford, Martin. *The Lights in the Tunnel: Automation, Accelerating Technology and the Economy of the Future*. Acculant, 2009.
———. *Rise of the Robots: Technology and the Threat of a Jobless Future*. New York: Basic, 2015.
Foster, John Bellamy. *The Vulnerable Planet: A Short Economic History of the Environment*. New York: Monthly Review, 1994.
Fox, Matthew. *Original Blessing*. Rochester, VT: Bear, 1983.
Frank, Andre Gunder. *Capitalism and Underdevelopment in Latin America*. New York: Monthly Review, 1967.
Freire, Paulo. *Pedagogy of the Oppressed*. New York: Continuum, 2005 [1970].
Frey Carl B., and Michael A. Osborne. "The Future of Employment: How Susceptible Are Jobs to Computerisation?" 2013. http://www.oxfordmartin.ox.ac.uk/downloads/academic/The_Future_of_Employment.pdf.
Freyfogle, Eric T., ed. *The New Agrarianism: Land, Culture, and the Community of Life*. Washington, DC: Island, 2001.
Friends of the Earth. "The Principles for the Oversight of Synthetic Biology." 2012. http://www.biosafety-info.net/file_dir/15148916274f6071c0e12ea.pdf.
Fuller, Steve, and Veronika Lipinska. *The Proactionary Imperative: A Foundation for Transhumanism*. New York: Palgrave Macmillan, 2014.
Fung, Archon, and Eric Olin Wright, eds. *Deepening Democracy: Institutional Innovations in Empowered Participatory Governance*. London: Verso, 2003.
Gamwell, Franklin. *The Divine Good: Modern Moral Theory and the Necessity of God*. San Francisco: HarperSanFrancisco, 1990.
Ganz, Marshall. "Leading Change: Leadership, Organization, and Social Movements." In *Handbook of Leadership Theory and Practice*, edited by Nitin Nohria and Rakesh Khurana, 527–68. Boston: Harvard Business, 2010.
Garcia-Roberts, Gus. "Five Reasons Turkey Point Could Be the Next Nuclear Disaster." *Miami New Times*, March 31, 2011.
Gibson-Graham, J. K. *A Postcapitalist Politics*. Minneapolis: University of Minnesota Press, 2006.
Giddens, Anthony. *Modernity and Self-Identity*. Malden, MA: Polity, 1991.
Gilding, Paul. *The Great Disruption*. New York: Bloomsbury, 2012.
Goldin, Claudia, and Lawrence F. Katz. *The Race between Education and Technology*. Cambridge, MA: Harvard University Press, 2009.
Goldin, Ian, and Mike Mariathasan. *The Butterfly Defect: How Globalization Creates Systemic Risk, and What to Do about It*. Princeton, NJ: Princeton University Press, 2014.
Gorz, Andre. *Reclaiming Work: Beyond the Wage-Based Society*. Malden, MA: Polity, 1999.
Gould, Kenneth A., et al. *The Treadmill of Production*. New York: Routledge, 2016.
Graeber, David. *Debt: The First 5,000 Years*. Brooklyn: Melville House, 2014.
Graham, Mark. "The Unsavory Gamble of Industrial Agriculture." In *Just Sustainability: Technology, Ecology, and Resource Extraction*, edited by Christiana Z. Peppard and Andrea Vicini, 105–16. Maryknoll, NY: Orbis, 2015.
Grant, Adam. *Give and Take*. New York: Viking, 2013.
Greeley, Andrew. *No Bigger Than Necessary: An Alternative to Socialism, Capitalism, and Anarchism*. New York: New American Library, 1977.

Groves, Christopher. "Living in Uncertainty: Anthropogenic Global Warming and the Limits of 'Risk Thinking.'" In *Future Ethics: Climate Change and Apocalyptic Imagination*, edited by Stefan Skrimshire, 118–26. New York: Continuum, 2010.
Guardini, Romano. *The Essential Guardini: An Anthology of the Writings of Romano Guardini*. Chicago: Liturgy Training, 1997.
Gutiérrez, Gustavo. *A Theology of Liberation: History, Politics and Salvation*. Translated by Sister Caridad Inda and John Eagleson. Maryknoll, NY: Orbis, 1973.
Hacker, Jacob. *The Great Risk Shift*. Oxford: Oxford University Press, 2008.
Haigerty, Leo J., ed. *Pope Pius XII and Technology*. Milwaukee: Bruce, 1962.
Haidt, Jonathan. *The Righteous Mind*. New York: Vintage, 2012.
Hajer, Maarten. *The Politics of Environmental Discourse*. Oxford: Clarendon, 1995.
Hardt, Michael, and Antonio Negri. *Multitude: War and Democracy in the Age of Empire*. New York: Penguin, 2004.
Harrison, Neil E. *Constructing Sustainable Development*. Albany: SUNY Press, 2000.
Hart, John. *Cosmic Commons: Spirit, Science, and Space*. Eugene, OR: Wipf and Stock, 2013.
———. *Ethics and Technology*. Cleveland: Pilgrim, 1997.
Haught, John. *Making Sense of Evolution*. Louisville: Westminster John Knox, 2010.
Hawk, L. Daniel. "The Truth about Conquest: Joshua as History, Narrative, and Scripture." *Interpretation: A Journal of Bible and Theology* 66, no. 2 (2012) 129–40.
Hawken, Paul, et al. *Natural Capitalism: Creating the Next Industrial Revolution*. Boston, MA: Little, Brown, 1999.
Hayes, Christopher. "The New Abolitionism." *The Nation*, April 22, 2014.
Hobgood, Mary E. *Catholic Social Teaching and Economic Theory: Paradigms in Conflict*. Philadelphia: Temple University Press, 1991.
Hock, Dee. *One From Many: VISA and the Rise of Chaordic Organization*. San Francisco: Barrett-Koehler, 2005.
Holland, Joe. *Modern Catholic Social Teaching: The Popes Confront the Industrial Age 1740–1958*. New York: Paulist, 2003.
Holt, Douglas B. "Why the Sustainable Economy Movement Hasn't Scaled: Toward a Strategy That Empowers Main Street." In *Sustainable Lifestyles and the Quest for Plenitude: Case Studies of the New Economy*, edited by Juliet B. Schor and Craig J. Thompson, 202–32. New Haven, CT: Yale University Press, 2014.
Hora, Roy. *The Landowners of the Argentine Pampas: A Social and Political History 1860–1945*. Oxford: Clarendon, 2001.
Howard, Philip N. *Pax Technica: How the Internet of Things May Set Us Free or Lock Us Up*. New Haven, CT: Yale University Press, 2015.
Huesemann, Michael, and Joyce Huesemann. *Techno-Fix: Why Technology Won't Save Us or the Environment*. Gabriola Island, BC: New Society, 2011.
Hughes, James. *Citizen Cyborg: Why Democratic Societies Must Respond to the Redesigned Human of the Future*. Cambridge, MA: Westview, 2004.
———. "A Strategic Opening for a Basic Income Guarantee in the Global Crisis Being Created by AI, Robots, Desktop Manufacturing and BioMedicine." *Journal of Evolution and Technology* 24, no. 1 (2014) 45–61.
Hughes, John. *The End of Work: Theological Critiques of Capitalism*. Oxford: Blackwell, 2007.

Hughes, Thomas P. "Technological Momentum." In *Does Technology Drive History? The Dilemma of Technological Determinism*, edited by Merritt Row Smith and Leo Marx, 101–13. Cambridge, MA: MIT Press, 1994.

Jackson, Tim. *Prosperity without Growth: Economics for a Finite Planet*. London: Earthscan, 2009.

Jacobson, Mark Z., and Mark A. Delucchi. "A Plan to Power 100 Percent of the Planet with Renewables." *Scientific American* (2009) 58–59.

———. "Providing All Global Energy with Wind, Water, and Solar Power, Part I: Technologies, Energy Resources, Quantities and Areas of Infrastructure, and Materials." *Energy Policy* 39 (2011) 1154–69, 1170–90.

Jenkins, Willis. *The Future of Ethics: Sustainability, Social Justice, and Religious Creativity*. Washington, DC: Georgetown University Press, 2013.

Jonas, Hans. *The Imperative of Responsibility: In Search of an Ethics for the Technological Age*. Chicago: University of Chicago Press, 1984.

Jordan, Bill. "Basic Income and the Common Good." In *Arguing for Basic Income: Ethical Foundations for a Radical Reform*, edited by Phillipe Van Parijs, 155–77. Brooklyn: Verso, 1992.

Kaufmann, Stuart. *At Home in the Universe: The Search for the Laws of Self-Organization and Complexity*. Oxford: Oxford University Press, 1996.

Kearney, Richard. *In the Wake of Imagination*. Minneapolis: University of Minnesota Press, 1988.

Kelly, Kevin. *What Technology Wants*. New York: Viking, 2010.

Keller, Catherine. *On the Mystery: Discerning Divinity in Process*. Minneapolis: Fortress, 2008.

Kellert, Stephen R., and E. O. Wilson. *The Biophilia Hypothesis*. Washington, DC: Island, 1995.

King, Martin Luther, Jr. *Where Do We Go from Here: Chaos or Community?* Boston: Beacon, 1968.

Klas, Mary Ellen. "FPL customers will be charged for fracking activities, board says." *Miami Herald*, June 18, 2015.

———. "Watchdog report says power companies wield too much influence in Florida Legislature." *Tampa Bay Times*, March 30, 2014.

Klein, Naomi. *This Changes Everything: Capitalism vs. The Climate*. New York: Simon & Schuster, 2103.

Kolbert, Elizabeth. *The Sixth Extinction: An Unnatural History*. New York: Holt, 2014.

Konczal, Mike. "Thinking Utopian: How about a Universal Basic Income?" *Washington Post*, May 11, 2013.

Krugman, Paul. "Sympathy for the Luddites." *New York Times*, June 13, 2013.

Kurzweil, Ray. *The Singularity Is Near*. New York, Viking, 2005.

Lanier, Jaron. *Who Owns the Future?* New York: Simon & Schuster, 2013.

Larkin, Amy. *Environmental Debt: The Hidden Costs of a Changing Global Economy*. New York: Palgrave, 2013.

Lamoureux, Patricia A. "Commentary on Laborem exercens (On Human Work)." In *Modern Catholic Social Teaching: Commentaries and Interpretations*, edited by Kenneth R. Himes, 399–412. Washington, DC: Georgetown University Press, 2005.

Laqueur, Thomas. "Some Damn Foolish Thing." *London Review of Books* 35, no. 23 (2013) 11–16.

Lanz, Tobias J., ed. *Beyond Capitalism & Socialism: A New Statement of an Old Ideal.* Norfolk, VA: IHS, 2008.
Lonergan, Bernard. *Insight: A Study of Human Understanding.* New York: Harper & Row, 1978.
Longman, Phillip. "Why the Economic Fates of America's Cities Diverged." *Atlantic,* November 28, 2015, 45–50.
Lovelock, James. *The Vanishing Face of Gaia: A Final Warning.* New York: Basic, 2009.
Luttig, Matthew. "The Structure of Inequality and Americans' Attitudes Towards Redistribution." *Public Opinion Quarterly* 77, no. 3 (2013) 811–21.
Mani, Anandi, et al. "Poverty Impedes Cognitive Function." *Science* 341, no. 6149 (2013) 976–80.
Manning, Erin. *Always More Than One: Individuation's Dance.* Durham, NC: Duke University Press, 2013.
Martin, Diarmund. "Nuclear Energy Must Be Used Not Only Peacefully But Safely." Address on September 22, 1998 to the 42nd General Conference of the International Atomic Energy Agency in Vienna, Austria.
Mason, Paul. *Postcapitalism: A Guide to Our Future.* New York: Farrar, Straus and Giroux, 2015.
Mauss, Marcel. *The Gift: The Form and Reason for Exchange in Archaic Societies.* Trans. W. D. Halls. New York: Norton, 2000.
Mayer, Jane. *Dark Money: The Hidden History of the Billionaires Behind the Rise of the Radical Right.* New York: Doubleday, 2016.
McDonough, William, and Michael Braungart. *The Upcycle: Beyond Sustainability—Designing for Abundance.* New York: North Point, 2013.
McGonigal, Jane. *Reality Is Broken: Why Games Make Us Better and How They Can Change the World.* New York: Penguin, 2011.
McKenzie, Jon. *Perform or Else: From Discipline to Performance.* New York, Routledge, 2001.
McKibben, Bill. *Eaarth: Making a Life on a Tough New Planet.* New York: St. Martin's, 2011.
McKinsey Global Institute. "Debt and (not much) Deleveraging." February 2015. http://www.mckinsey.com/global-themes/employment-and-growth/debt-and-not-much-deleveraging.
———. "Disruptive technologies: Advances that will transform life, business, and the global economy." May 2013. http://www.mckinsey.com/business-functions/digital-mckinsey/our-insights/disruptive-technologies.
———. "The world at work: Jobs, pay, and skills for 3.5 billion people." June 2012. http://www.mckinsey.com/global-themes/employment-and-growth/the-world-at-work.
McNeill, J. R., and Peter Engelke. *The Great Acceleration: An Environmental History of the Anthropocene since 1945.* Cambridge, MA: Belknap, 2014.
Meadows, Donella H., et al. *Beyond the Limits: Confronting Global Collapse, Envisioning a Sustainable Future.* White River Junction, VT: Chelsea Green, 1992.
Merritt, Carolyn. *Tango Nuevo.* Gainsville: University Press of Florida, 2011.
Milbank, John. *Theology and Social Theory: Beyond Secular Reason.* 2d ed. Malden, MA: Blackwell, 2006.
Milbank, John, and Adrian Pabst. *The Politics of Virtue: Post-Liberalism and the Human Future.* London: Rowman & Littlefield, 2016.

Miller, Vincent J. *Consuming Religion: Christian Faith and Practice in a Consumer Culture*. New York: Continuum, 2003.

———. "History Can Guide the Future of Catholic Social Teaching." *ACCU Update* (2012). Accessed September 10, 2015. http://www.accunet.org/i4a/pages/index.cfm?pageID=3754.

Misner, Paul. *Social Catholicism in Europe*. New York: Crossroad, 1991.

Mokyr, Joel, et al. "The History of Technological Anxiety and the Future of Economic Growth: Is This Time Different?" *Journal of Economic Perspectives* 29, no. 3 (2015) 31–50.

Mol, Arthur P., and David A. Sonnenfeld. "Ecological Modernisation Around the World: An Introduction." In *Ecological Modernisation Around the World*, edited by Arthur P. Mol and David A. Sonnenfeld, 3–12. London: Cass, 2000.

Moltmann, Jürgen. *Theology of Hope*. Minneapolis: Fortress, 1993.

Moore, Jason W. *Capitalism in the Web of Life*. London: Verso, 2015.

More, Max. "The Proactionary Principle." 2005. http://www.maxmore.com/proactionary.html.

Morris, Simon Conway. *The Runes of Evolution: How the Universe Became Self-Aware*. West Conshohoken, PA: Templeton, 2015.

Morrison, Elizabeth Wolfe, and Frances J. Milliken. "Organizational Silence: A Barrier to Change and Development in a Pluralistic World." *Academy of Management Review* 25, no. 4 (2000) 706–725.

Moylan, Tom. "Bloch against Bloch: The Theological Reception of *Da Prinzip Hoffnung* and the Liberation of the Utopian Function." In *Not Yet: Reconsidering Ernst Bloch*, edited by Jamie Owen Daniel and Tom Moylan, 96–121. New York: Verso, 1997.

Mykleby, Mark, et al. *The New Grand Strategy: Restoring America's Prosperity, Security, and Sustainability in the 21st Century*. New York: St. Martin's, 2016.

Noble, David R. *The Religion of Technology: The Divinity of Man and the Spirit of Invention*. New York: Knopf, 1997.

Northcott, Michael S. *Place, Ecology and the Sacred*. London: Bloomsbury, 2015.

Norton, Bryan G. *Sustainability: A Philosophy of Adaptive Ecosystem Management*. Chicago: University of Chicago Press, 2005.

Nuclear Regulatory Commission. "Environmental Impact Statement for Combined Licenses (COLs) for Turkey Point Nuclear Plant Units 6 and 7." *Final Report for Comment* 1 (October 2016).

Nunez, Theodore W. "The Ecological Common Good: Responding to the Problem of Urban Sprawl." In *Ethical Dilemmas in the Third Millenniu*, edited by Francis Eigo, 2:1–36. Proceedings of the Villanova Theology Institute 33. Villanova, PA: Villanova University Press, 2001.

———. "Rolston, Lonergan and the Intrinsic Value of Nature." *Journal of Religious Ethics* 27, no. 1 (1999) 105–28.

O'Brien, Mary H. *Making Better Environmental Decisions: An Alternative to Risk Assessment*. Cambridge, MA: MIT Press, 2000.

O'Connor, James. *Natural Causes: Essays in Ecological Marxism*. New York: Guilford, 1997.

O'Meara, Thomas F. *Vast Universe: Extraterrestrials and Christian Revelation*. Collegeville, MN: Liturgical, 2012.

Ophuls, William. *Requiem for Modern Politics: The Tragedy of the Enlightenment and the Challenge of the New Millennium*. Boulder, CO: Westview, 1998.

Oreskes, Naomi, and Erik M. Conway. *Merchants of Doubt: How a Handful of Scientists Obscured the Truth on Issues from Tobacco Smoke to Global Warming.* New York: Bloomsbury, 2010.
Perrow, Charles. *Normal Accidents: Living with High-Risk Technologies.* Princeton, NJ: Princeton University Press, 1999.
Perry, William J. *My Journey at the Nuclear Brink.* Stanford: Stanford University Press, 2015.
Pew Research Center National Survey. "Most Say Government Policies Since Recession Have Done Little to Help Middle Class, Poor." February 2015. http://www.people-press.org/2015/03/04/most-say-government-policies-since-recession-have-done-little-to-help-middle-class-poor/.
Phelps, Edmund S. *Mass Flourishing: How Grassroots Innovation Created Jobs, Challenge, and Change.* Princeton, NJ: Princeton University Press, 2013.
Piketty, Thomas. *Capital in the Twenty-First Century.* Cambridge, MA: Belknap, 2014.
Polanyi, Karl. *The Great Transformation: The Political and Economic Origins of Our Time.* Boston: Beacon, 2001.
Postman, Neil. *Technopoly: The Surrender of Culture to Technology.* New York: Knopf, 1992.
Prigogine, Ilya, and Isabelle Stenger. *Order Out of Chaos.* New York: Bantam, 1984.
Redclift, Michael, and Ted Benton, "Introduction." In *Social Theory and the Global Environment*, edited by Michael Redclift and Ted Benton, 1–14. London: Routledge, 1994.
Reich, Robert. *Saving Capitalism: For the Many, Not the Few.* New York: Knopf, 2015.
Rifkin, Jeremy. *The End of Work: The Decline of the Global Labor Force and the Dawn of the Post-Market Era.* New York: Putnam, 1995.
———. *The Third Industrial Revolution: How Lateral Power Is Transforming Energy, the Economy, and the World.* New York: St. Martin's, 2011.
———. *The Zero Marginal Cost Society.* New York: Palgrave Macmillan, 2014.
Rohde, David. "The Swelling Middle." 2012. http://www.reuters.com/middle-class-infographic.
Rolston, Holmes. *Conserving Natural Value.* New York: Columbia University Press, 1994.
———. *Environmental Ethics: Duties to and Values in the Natural World.* Philadelphia: Temple University Press, 1989.
———. "Is There an Ecological Ethic?" *Ethics* 85, no. 2 (1975) 93–105.
———. *Science and Religion: A Critical Survey.* Philadelphia: Temple University Press, 1987.
Rosa, Hartmut. *Social Acceleration: A New Theory of Modernity.* New York: Columbia University Press, 2013.
Roth, Kenneth M. *Annihilating Difference: The Anthropology of Genocide.* Berkeley, CA: University of California Press, 2002.
Rushkoff, Daniel. *Present Shock: When Everything Happens Now.* New York: Penguin, 2013.
———. *Throwing Rocks at the Google Bus.* New York: Penguin, 2016.
Sachs, Jeffrey D., and Laurence J. Kotlikoff. "Smart Machines and Long-Term Misery." NBER Working Paper No. 18629. December 2012. http://www.columbia.edu/~lnp3/jeffrey_sachs.pdf.

Scheiber, Noam. "Rising Economic Insecurity Tied to Decades-Long Trend in Employment Practices." *New York Times*, July 12, 2015.

Schell, Jonathan. *The Unconquerable World: Power, Nonviolence, and the Will of the People*. New York: Holt, 2003.

Scheuerman, William. "Citizenship and Speed." In *High-Speed Society: Social Acceleration, Power, and Modernity*, edited by Hartmut Rosa and William Scheuerman, 297–316. University Park, PA: Pennsylvania State University Press, 2009.

Schlosberg, David. *Defining Environmental Justice: Theories, Movements, and Nature*. Oxford: Oxford University Press, 2009.

Schneider, Nathan. "Truly, Much Can Be Done!: Cooperative Economics from the Book of Acts to Pope Francis." In *Laudato Si': Ethical, Legal, and Political Implications*, edited by Frank Pasquale, 242–63. Cambridge: Cambridge University Press, 2017.

Schor, Juliet B. *Plenitude: The New Economics of True Wealth*. New York: Penguin, 2010.

Schor Juliet B., and Craig J. Thompson, eds. *Sustainable Lifestyles and the Quest for Plenitude: Case Studies of the New Economy*. New Haven, CT: Yale University Press, 2014.

Senge, Peter M., et al. *Presence: Human Purpose and the Field of the Future*. New York: Doubleday, 2004.

Shapiro, Judith. *Mao's War Against Nature: Politics and the Environment in Revolutionary China*. Cambridge: Cambridge University Press, 2001.

Shiva, Vandana. *Making Peace with the Earth*. London, Pluto, 2013.

Shneier, Bruce. *Data and Goliath*. New York: Norton, 2015.

Shutkin, William. *The Land That Could Be: Environmentalism and Democracy in the Twenty-First Century*. Cambridge, MA: MIT Press, 2000.

Simon, Julian L. *Ultimate Resource 2*. Princeton, NJ: Princeton University Press, 1996.

Simons, Robert G. *Competing Gospels: Public Theology and Economic Theory*. London: Dwyer, 1995.

Smith, Graham. *Democratic Innovations: Designing Institutions for Citizen Participation*. Cambridge: Cambridge University Press, 2009.

Southeast Florida Regional Climate Change Compact. "A Region Responds to a Changing Climate." October 2012. http://www.southeastfloridaclimatecompact.org/wp-content/uploads/2014/09/regional-climate-action-plan-final-ada-compliant.pdf.

Staletovich, Jenny. "Mayors make case against FPL nuclear expansion." *Miami Herald*, April 15, 2015.

———. "State eases oversight of Turkey Point cooling canals." *Miami Herald*, January 30, 2015.

Standing, Guy. *The Precariat: The New Dangerous Class*. London: Bloomsbury, 2011.

Steffen, Will, et al. "The Anthropocene: Are Humans Now Overwhelming the Great Forces of Nature?" *Ambio* 36, no. 8 (2007) 614–21.

Steffen, Will, et al. "Planetary boundaries: Guiding human development on a changing planet." *Science* 347, no. 6223 (2015) 737–46.

Stern, Andy. *Raising the Floor*. New York: Perseus, 2016.

Stone, Andrea. "8 Surprising, Depressing, and Hopeful Findings from Global Survey of Environmental Attitudes." *National Geographic*, September 27, 2014. http://news.nationalgeographic.com/news/2014/09/140926-greendex-national-geographic-survey-environmental-attitudes/.

Strate, Lance. *Amazing Ourselves to Death: Neil Postman's Brave New World Revisited.* New York: Lang, 2014.
Sudbury, Julia, ed. *Global Lockdown: Race, Gender, and the Prison-Industrial Complex.* New York: Routledge, 2004.
Taleb, Nassim N. *The Black Swan: The Impact of the Highly Improbable.* 2d ed. New York: Random House, 2010.
Tatum, Jesse. "Design and the Reform of Technology: Venturing Out into the Open." In *Technology and the Good Life?*, edited by Eric Higgs et al., 182–94. Chicago: University of Chicago Press, 2000.
Taylor, Sarah McFarland. *Green Sisters.* Cambridge, MA: Harvard University Press, 2007.
Thompson, Paul B. "Farming as Focal Practice." In *Technology and the Good Life?* edited by Eric Higgs et al., 166–81. Chicago: University of Chicago Press, 2000.
Thompson, Robert Farris. *Tango: The Art History of Love.* New York: Vintage, 2005.
Tracy, David. *The Achievement of Bernard Lonergan.* New York: Herder and Herder, 1970.
Turkle, Sherry. *Alone Together: Why We Expect More from Technology and Less from Each Other.* New York: Basic, 2011.
Turner, Bryan S., and Chris Rojek. *Society and Culture: Scarcity and Solidarity.* London: SAGE, 2001.
Tyson, Laura, and Michael Spence. "Exploring the Effects of Technology on Income and Wealth Inequality." In *After Picketty: The Agenda for Economics and Inequality*, edited by Heather Boushy et al., 170–208. Cambridge. MA: Harvard University Press, 2017.
University of Wisconsin Center for Cooperatives. *Research on the Economic Impact of Cooperatives.* 2014. http://reic.uwcc.wisc.edu/summary/.
U. S. Presidential Commission for the Study of Bioethical Issues. "New Directions: The Ethics of Synthetic Biology and Emerging Technologies." 2010. http://bioethics.gov/synthetic-biology-report.
Vague, Richard. "The Private Debt Crisis." *Democracy: A Journal of Ideas* (2016) 70–84.
Van Parijs, Philippe. *Real Freedom for All.* Oxford: Oxford University Press, 1997.
Van Parijs, Philippe, and Yannick Vanderborgt, *Basic Income: A Radical Proposal for a Free Society.* Cambridge, MA: Harvard University Press, 2017.
Van Wieren, Gretel. *Restored to Earth: Christianity, Environmental Ethics, and Ecological Restoration.* Washington, DC: Georgetown University Press, 2013.
Venter, J. Craig. *Life at the Speed of Light.* New York: Viking, 2013.
Vogel, David. *The Market for Virtue: The Potential and Limits of Corporate Social Responsibility.* Washington, DC: Brookings, 2005.
Wagner, Gernot, and Martin L. Weitzman. *Climate Shock: The Economic Consequences of a Hotter Planet.* Princeton, NJ: Princeton University Press, 2015.
Wahl, Daniel C. *Designing Regenerative Cultures.* Aberdour, UK: Triarchy, 2016.
Wajcman, Judy. *Pressed for Time: The Acceleration of Life in Digital Capitalism.* Chicago: University of Chicago Press, 2015.
Waldrop, M. Mitchell. "The Trillion-Dollar Vision of Dee Hock." *Fast Company* (October 1996).
Walker, Gordon. *Environmental Justice: Concepts, Evidence and Politics.* New York: Routledge, 2012.

Walker, Mark. "BIG and Technological Unemployment: Chicken Little versus the Economists." *Journal of Evolution and Technology* 24, no. 1 (2014) 5–25.
Wallerstein, Immanuel. *The Capitalist World-Economy*. London: Cambridge University Press, 1979.
———. *The End of the World As We Know It*. Minneapolis: University of Minnesota Press, 1999.
Weeks, Kathi. "Imagining Non-Work." *Social Text*. March 28, 2013. http://socialtextjournal.org/periscope_article/imagining-non-work-2/.
———. *The Problem with Work*. Durham, NC: Duke University Press, 2011.
Weisberg, Jacob. "We are Hopelessly Hooked." *New York Review of Books*, February 25, 2016, 37–40.
Weld, Virginia. *The Ethics of Care: Personal, Political, and Global*. Oxford: Oxford University Press, 2006.
White, Lynn, Jr. "The Historical Roots of Our Ecologic Crisis." *Science* 155 (1967) 1203–7.
Whitehead, Alfred North. *Science and the Modern World*. New York: Free, 1967.
Whiteside, Kerry H. *Precautionary Politics: Principle and Practice in Confronting Environmental Risk*. Cambridge, MA: MIT Press, 2006.
Widerquist, Karl, et al., eds. *Basic Income: An Anthology of Contemporary Research*. Malden, MA: Wiley Blackwell, 2013.
Wilkinson, Richard, and Kate Pickett. *The Spirit Level: Why Greater Equality Makes Societies Stronger*. New York: Bloomsbury, 2010.
Wilson Center. "The DNA of the U.S. Regulatory System: Are We Getting It Right for Synthetic Biology?" October 2015. http://www.synbioproject.org/site/assets/files/1388/synbio_reg_report_final.pdf.
Wilson, William Julius. *When Work Disappears: The World of the New Urban Poor*. New York: Vintage, 1997.
Winner, Langdon. *Autonomous Technology: Technics-out-of-Control as a Theme in Political Thought*. Cambridge, MA: MIT Press, 1977.
Wogaman, Philip. *Guaranteed Annual Income: The Moral Issues*. Nashville: Abingdon, 1968.
World Economic Forum. *Global Risks 2013*. http://www3.weforum.org/docs/WEF_GlobalRisks_Report_2013.pdf.
Worster, Donald. *Nature's Economy: A History of Ecological Ideas*. Cambridge: Cambridge University Press, 1994.
Wright, Eric Olin. *Envisioning Real Utopias*. New York: Verso, 2010.
Wright, Richard T. *Myths America Lives By*. Champaign, IL: University of Illinois Press, 2004.
Zakaria, Fareed. *The Post-American World: Release 2.0*. New York: Norton, 2011.
Zamagni, Stefano. "Catholic Social Thought, Civil Economy, and the Spirit of Capitalism." In *The True Wealth of Nations: Catholic Social Thought and Economic Life*, edited by Daniel K. Finn, 63–93. Oxford: Oxford University Press, 2010.

Index

A
abundance, 4–5, 179; earth's, 44; sustaining material, 28
abuse, domestic, 167
acceleration: cultural, 23, 28, 189; promise of, 25; social, 7, 9, 22–26, 30, 33, 36, 109, 123, 221; technological, 23
accidents, 16–17, 85, 87, 194, 198; normal, 18
accountability, 6, 143, 189
acid rain, 71, 75
Ackerman, Diane, 183
action: co-creative, 133; devaluing Arendtian, 186; direct, 35, 40, 87, 155; far-sighted, 34; nonviolent, 38; political, 156; rational, 86; responsible, 129, 209; social, 11, 62, 73–74, 94
activists, 18, 116, 140, 147, 155, 178; climate-justice, 155; environmental, 149; environmental-justice, vii, 34; first-nation, 6; frontline, 155; labor rights, 166
activities: civic, 226; civil-economic, 81; countercultural, 123; daily, 35; global cyber-economic, 143; high-carbon, 173; industrial, 16; low cultural, 94; meaningful, 185; modern techno-scientific, 55; non-labor, 174; political, 52, 180; technological, 54

actors, 20, 119, 201; civil-society, 7, 19, 21, 137, 189; governmental, 19; powerful economic, 137; social, 29, 52
Adam, Barbara, 6–8, 11, 17, 29, 86
Adamic perfection, 25
adaptation, 139; climate-change, 193; systemic, 39
Adirondacks, 6
adventure, 130; ongoing, 201
affirmation, 52, 106, 127; clear, 73; dual, 88; explicit, 97; ongoing, 88; theologically-grounded, 98
affluence, 30, 63, 71, 102–3, 105–6, 117, 154, 156, 178; rising, 102; technological, 102; unsustainable, 156
Africa, 121; sub-Saharan, 74
African dance traditions, 222
age: digital, 102, 149; fast-paced technological, 116; machine, 162–64, 168; modern, 64, 185
agencies, 52, 80, 124, 202; federal investigative, 194; international, 18; local/indigenous, 202
agenda, 107; alternative political, 154; corporate-dominated, 35; ecological-modernization, 140; eco-modernizing, 73, 87; environmental, 71; nuclear-powered growth, 195; progressive eco-modern, 34; sustainable-growth, 72

249

250 INDEX

aggregate levels, 31
Agreement, Paris, 34, 39, 171

agri-business, 139
agriculture, 32, 48, 51; sustainable, 98, 107, 148
agro-ecology, 81, 145
algae blooms, 191
alienation, 9, 59, 62, 75, 123, 128, 183–84, 221
analyses: cost-benefit, 110; historical, 15, 57; orthodox, 52; risk-benefit, 32; scientific, 131; structural, 38–39, 57, 71, 74, 77, 79; systemic risk, 139
Anthropocene, 5, 35, 74, 196, 199, 219
anthropocentric, 55, 71, 79; bias, 200; hubris, 84; modern, 55, 83, 106
anthropology, theological, 55, 66, 160
Appalachia, 70, 78–80
apprenticeships, 170, 225
approach: consensus-building, 62, 66; dialogical, 95; eco-modernizing, 78, 87; entrepreneurial, 89; forward-looking, 111, 117; green-communitarian, 98; just-sustainability, 79; market-oriented, 30; multi-disciplinary, 8; planetary-management, 35; problem-based, 96; social-learning, 40; technocratic, 55; techno-progressive, 36; top-down, 72, 94; transformational, 40
Aquinas, Thomas, 44–45, 55
Arab Spring, 21
Arendt, Hannah, 185–86
Argentina, 222, 226
Aristophanes, 127
artificial intelligence, x–xi, 162; useful, 163
artisans, 58, 101; aspiring, 177
artistry, 8, 225, 235; refined, 228

artists, 26, 46–47, 147, 203
Asian financial crisis, 16
aspirations, 21, 58, 74, 92, 124; utopian, 127
assumptions: background, 29; basic, 52; early-modern, 17, 34, 76, 85–86; historicist, 57; mechanistic, 95; modern, 34; normative, 52; secular-modernist, 45; state-socialist, 46; technocratic, 86
Australia, 79
authoritarian: governments, 197; rule, 126, 157, 176
authority: decision-making, 97; legal, 19, 189; public, 53, 59; true world political, 98
automation, 3, 49–50, 101, 106, 139–40, 159, 161–67, 169–71, 179, 182, 185, 187, 216
autonomy, 142, 174, 216, 218; limited, 54; technological, 85

B
balance, 20, 37, 70, 83, 147, 162, 199; dynamic, 82, 150, 209; work-life, 54, 186
bankruptcy, 24, 194
banks, 20, 144, 203; local, 148
Barbour, Ian, 11
basic income, viii, 144–45, 149, 154, 161, 169, 171–77, 180, 187; universal, xi, 94, 144, 175
Beck, Ulrich, 5, 7, 16–19, 21, 38, 62, 75–76, 81, 85–86, 168, 181, 195–97
Benedict XVI (Pope), 11, 72, 91, 182, 213
bio-conservatives, 10, 28–29, 32, 190, 213
biology, synthetic, 3, 23, 189, 205–6, 208
biosphere, 9, 38, 97, 151, 196; integrity, 9, 91
black swan blindness, 20, 189, 205
Bloch, Ernst, 115, 125

Borgmann, Albert, 9, 11, 100–106, 117, 123, 146, 148, 160, 183, 186, 214, 234
BP (British Petroleum), 194

C
capital accumulation, 4, 24–25, 38
capitalism, 1, 39, 45, 57, 102, 152, 235; global, 38, 77, 156; historical, 37, industrial, 43, 46, 59, 64; late, 26, 104; liberal, 37, 44, 46, 72; sustainable, 33, 38; technocratic, 11, 59
capitalist, 24–26, 32, 34, 36–37, 40, 43, 46, 57, 59, 80, 85, 123–24, 150, 182–83, 187
carbon dioxide, 7, 76
carbon dividends, 171, 173–74
Caritas in veritate, 92–93, 182
Catholicism, social, 10, 42–45, 147
Centesimus annus, 59, 76
Christian leaders, 10–11, 111, 171, 187
Christian living, authentic, 87, 117, 130–31
Christians, ix–x, 3, 10, 49, 53, 60, 88, 92, 107, 117, 128, 130–31, 156, 159, 175–76
church teaching, 42–43, 50, 65–66, 69, 83, 106, 182
citizens, ordinary, 7, 17, 20–21, 118, 126, 189
citizenship, 27, 176
civilization, 21, 47, 198
climate change, 7, 9, 16–20, 68, 91, 95, 98, 134, 138–39, 146, 148, 168–69, 171, 187, 192–95
Cold War, 6, 52, 71
collaboration, 40, 98–99, 120, 140, 155, 160, 206
communitarian ethos, 46, 50, 67, 69, 77, 79, 147; green, 29, 34–39, 62, 68–69, 73–74, 81, 98, 207

community: local, 68, 80, 146, 148, 178, 195, 197, 202; self-development, 51, 145
consumerism, 26, 64, 77, 83, 107–8
consumption, 3–4, 9, 24, 33, 63, 101, 103–4, 143, 155, 170, 187, 234
contemplation, 75, 107, 147, 161, 184
cooperatives, worker-owned, 44, 47, 90, 121–22
corporations, transnational, 53, 59, 80
corruption, 40, 54, 118
countries: advanced-industrial, 21, 61, 93; developing, 3, 31, 99, 145, 164, 167; poor, 30, 51; highly-indebted, 68; indebted, 137; Protestant, 224
creative freedom, 117, 122–23
creative middle, 188–89, 191, 193, 195, 197, 199, 201, 203, 205, 207–9, 211, 213
culture: democratic, 144, 184; dialectic of, 211–12

D
debt, 7, 137, 139, 155, 177
democracy, 24, 42, 50, 117, 155, 197; deliberative, 7, 27, 189; direct, 148, 154; participatory, 9, 35, 38, 118, 126, 186
democratic control, 80, 118, 154
democratization, 37, 48, 78
deregulation, 30–31
desertification, 18, 97
development, 50, 52, 54–56, 58, 68, 70–71, 73, 76–77, 82–83, 85, 91–92, 105–6, 192, 194, 207; human, 56, 64, 109, 129, 154; authentic, 43, 56, 63, 71, 201; sustainable, 109
devices, technological, 101–2, 105–6, 146, 219
dialectics, integral, 128–29, 209, 211–12

digitalization, 3, 142, 163, 216
dignity, human, 46, 52, 81
distributists, 47, 147–48; British, 43
divine, 128, 184, 200, 211;
    intelligence, 135, 150;
    mystery, 131, 184, 235
domination, neocolonial, 51, 61
dominion, 49, 51, 61, 63, 66, 70–72
Doomsday Clock, 6
Durbin, Paul, 11

E
ecological: economists, 9, 34, 38;
    modernization, 32, 38, 40,
    72, 78; paradigm, 86, 88, 91,
    95–96, 99, 107
ecology, 10, 42, 59, 69, 79, 83, 85,
    95, 97, 99, 195, 202, 229;
    human, 59–60, 99
eco-modernizers, 34, 36, 38, 40–41,
    69, 73, 87; progressive,
    39–40
economic: growth, 9, 15–16, 31,
    35, 52, 55, 97, 156, 171;
    opportunity, 44, 48, 78, 81,
    118, 142, 146, 174, 176–77;
    practices, 83, 86, 92, 94, 97,
    162
economy: clean-tech, 7, 99; hybrid,
    40, 91, 93, 136, 140, 146,
    152, 154, 156, 177, 179–80;
    market, civil, 90, 93–94, 140,
    149, 178; steady-state, 9,
    38, 170
eco-social question, 1, 5, 8, 11, 65,
    77, 79, 95, 136, 160, 170, 214
ecosystems, 71–72, 97–98, 149, 170,
    200, 204
EIS (environmental impact
    statement), 192
elections, 119, 148
elites, 7, 17, 33, 35, 66, 72, 77, 87, 94,
    98, 155, 166, 210
empathy, 9, 97, 116, 122, 130, 197,
    234–35
employment, 53–54, 64, 94, 138,
    161–64, 168, 175, 180, 182;
    formal, 145, 163, 167, 172,
    181
energy: nuclear, 2, 76; options,
    76, 190, 204; production,
    148–49, 155; renewable, 32,
    120–21, 148, 152, 164
Enlightenment, 42, 45, 101
environment, hybridized, 35
environmental: crisis, 37, 50, 59–60,
    64, 66, 70, 75, 82, 84, 99;
    degradation, 38, 60, 69,
    72, 74, 80, 85, 99, 157, 159;
    refugees, 18, 216; risks, 9,
    16–17, 70, 81, 85, 110, 139,
    189
Epicurus, 46
ethics: global, 82, 84, 199, 202;
    human rights, 52
European economies, 48, 222
Evangelii Gaudium, 62
evolution, 133, 198–201, 222;
    biological, 109, 199; natural,
    198–99; technological, 199
exploitation, neocolonial, 78, 85

F
family life, 161, 186, 231
farming, vertical, 120, 145–46
Ferre, Frederick, 11
Francis (Pope), 11, 62, 86, 91,
    95–99, 105–11, 123, 160,
    182, 213
Francis (Saint), 8
freedom: political, 142, 181, 184;
    substantive, 147, 172,
    180–81
Fritsch, Albert, 11
fuel, fossil, 145, 155, 164
funding, public, 100, 148, 179
futurists, Christian, x–xi

G
Gandhi, Mahatma, 35
Gates, Bill, 40
GDP (Gross Domestic Product),
    56, 173; growth, 154, 161;
    World, 3–4, 90, 137

globalization, 24, 51–52, 66, 71, 79, 81, 99, 136–37; economic, 33, 90
global North, 61, 71–72, 78, 98, 116
global society, 95, 109, 138, 157, 163, 189
global South, 3, 18, 71–72, 74, 78, 93, 98, 116, 121, 139, 141, 144–45, 165, 182, 208
Gore, Al, 40
governance, 28, 33–35, 40, 154
gratitude, 8, 83, 92, 104, 115–16, 129–30, 184, 218–19, 233, 235
gratuitousness, principle of, 91–94
Great Recession, 20, 91

H
happiness, 25–26, 102
Hart, John, 11, 125, 134, 201
Hawking, Stephen, 162
hazards, 85–86; industrial, 16; invisible, 16; moral, ix
healing, 8, 129, 212
health: emotional, 160; environmental, 19, 31, 75, 84, 110, 191, 205, 207; psychic, 186; public, 84, 118, 193–94, 212
health care, 90, 137, 146, 148, 163, 172; universal, 176, 180
heaven and earth, new, 130
Hebrew prophets, ix
Hefner, Philip, 11
hegemony, cultural, 186, 235
high-speed society, 24, 211
Hiroshima, 6
history, ix–x, 5–6, 8, 20–21, 23, 25, 78–79, 125, 127, 130, 132, 163, 189–90, 209, 211–12; colonial, 4; human, 111, 129, 211; modern, 162; social, 199, 221, 223
Hock, Dee, 135, 150–52, 158, 212
Hoover Dam Bridge, 2
horizon, 63, 125, 131, 139, 159, 182, 202; alternative, 125, 161; limitless, 51; longer-term, 24; new, 60; transcendent, 25, 55
hospitality, generous, 231
housing, 144, 172; affordable, 99; public, 153
Hughes, James, 172
human capacities, 85, 125, 129; evolved, 122
humanism: authentic, 99; authentic religious, 55; integral, 55; secular, 55
humility, 8, 11, 116, 130, 219; epistemic, 207; false, 84
hurricanes, 190, 193
hydraulic fracturing, 204
hyperacceleration: current phase of, 24–25; late-modern, 24; phase of, 102, 189

I
Iceland, 119, 126
identities, 15, 21, 29, 167, 181–82; national, 226; new, 179; shared, 108
ideological, 17, 27; bias, 206; conservatism, 170; distortions, 42; excesses, 67; left-right, 28
ideologies, ix, 46, 56, 124, 131–32; communist, 128; free-market, 77; modern, 147; narrow, 99; neo-liberal, 72
IMF (International Monetary Fund), 90; riots, 68
imagination, 1, 9, 43, 45–46, 66–67, 69, 115–16, 125, 132–33, 140, 144, 151, 158, 229, 234; arcadian, 36, 69, 74–75, 79; collective, 20; communitarian, 45; concrete-utopian, 126–28, 131, 154; enlightened technocratic, 33; ethical, 46, 67; ethical-poetical, 116, 126; forward-looking, 131; metaphorical, 29; poetical, 67; poetical-ethical, 67; prophetic-utopian, 58, 67,

69, 81, 116, 124, 131, 136, 158; romantic, 46; utopian, 78
immigrants, 138, 222; working-class, 170
improvisation, 223, 227
incentives, 63, 181; non-material, 94
inclusion, 119; social, 77, 96, 154, 234
income, 15, 169–70, 174; additional, 183; disposable, 173; national, 15, 226
India, 116, 145, 165; ancient, 22; rural, 74
individualism, 197; expressive, 47; selfish, 64
individuality, 135, 150, 210
industrial: agriculture, 148; chemicals, 208; era, 207; modernity, 18–19, 40, 60–61, 64, 66, 70, 76, 81, 197; pollutants, 74; risks, 16–17; society, 15–16, 35, 147, 177, 197; wastes, 80
industries, 3, 17, 24, 27, 32, 37, 44, 48, 163–65, 167, 195; corporate-culture, 105; labor-intensive, 106; modern, 47; new, 162; nuclear power, 18
inequality: economic, x; extreme social, 57; global, 71, 137, 168; growing, 22, 71, 90, 161, 170; reducing, 175; rising, 15, 159, 162, 166–67, 170–71, 180; widening, 30
information, 23–24, 95–96, 108, 142–43, 149, 194, 206
infrastructure: new, 141; smart, 142, 144, 146, 149, 177
injustice: environmental, 73; historic, 176; present, 130; structural, 58, 66; systemic, 37, 57, 64, 71, 77
innovation, 23–25, 28, 32–33, 35, 37, 90–91, 94, 97, 120, 123, 148–50, 153–54, 162–63, 178–79, 205–6; indigenous,

148, 178; open-source, 90; private-sector, 105; scientific, 86; social, 35, 91, 94, 153, 158, 178–79
insecurity, 21–22, 124, 147, 167; economic, 22, 177
instability, 176; chronic, 22; political, 138–39, 176
instincts, 49; evolved social, 122, 197
institutions, 34, 39, 87, 93, 118, 123, 153–54, 157, 161, 173, 178, 180, 204, 209, 212; alternative, 40, 153, 180; cultural, 94; democratic, 197; historic, 21; modern, 24; political, 187; reinventing, 91, 111, 122; religious, 105; social, 33; supportive, 123
insurance, 16, 122; private health, 22
intelligence, 51, 61, 82, 111
interdependence, 232; growing, 51
interests: commercial, 208; economic, 37, 109, 195; elite, 107; entrenched, 155; financial, 93; mutual, 56; powerful agro-industrial, 82; powerful bio-technology, 213; vested, 194
internet of things. *See* IoT
intimacy, 217, 219–20, 228
invention, 3, 67, 163, 198, 219; cultural, 198; modern, 2; new, 162, 206, 223
investment, 57, 77, 80, 137, 167, 174; corporate, 121; foreign direct, 3; strategies, 174
IoT (internet of things), xi, 3, 23, 102, 141–42, 144, 163, 167; platform, 141; users, 143
ISIS, 145
isolation, 95, 99, 202, 234

J
Japan, 32, 116, 137–38
Jesuits, 226

jobs, ix, 3, 161–62, 165, 168, 172, 180–81, 204, 216; dead-end, 80; full-time, 179; green, 164; living-wage, 121; low-paying, 178; non-routine, 165; soul-deadening, 46; white-collar, 165
John Paul II (Pope), 50, 59, 62–66, 69–72, 76–77, 87, 99, 182, 213, 217
John XXIII (Pope), 48, 51–52, 56, 61
Jonas, Hans, 196–97, 204, 213
joy, 92, 104, 108, 184, 218–19, 223, 233–35
judgment, 28, 105, 208; informed, 19; normative, 96; objective, 76; prudential, 87; religious, 130
justice, 18, 29, 51–52, 77–78, 81, 92–93, 95, 129–30, 155, 206, 234; absolute, 62; distributive, 46, 52, 174, 180; economic, 42, 168, 176–77; environmental, 18, 105, 192; global, 72; intergenerational, 35; social, 50, 64–66, 73, 78, 123

K
Kelly, Kevin, 198–99, 201, 203–5, 209
Keystone XL Pipeline, 6
Klein, Naomi, 121, 149, 155
knowledge, 49, 54, 95, 106, 108, 143, 149, 204; local, 145, 203; perspectival, 29; scientific, 54
Kurzweil, Ray, 23, 189

L
labor, 23–24, 37, 39, 45, 53, 61, 63–64, 79, 100–101, 162, 164, 174–76, 182, 185, 233; cheap, 57, 164, 222; dignified, 53, 183; manual, 101, 165; mindless, 9, 101, 103, 186; organized, 180; productive, 63; self-organized, 181
Laborem exercens, 63-65, 72-73, 77, 182
laborers, 44, 185, 222; migrant, 22; unskilled, 101
land: agricultural, 176; arable, 99, 139; cheap, 79; freed, 45; promised, 5; trusts, 147, 153
land ownership, 80, 139, 154, 208; concentrated, 80; cooperative, 81
landscapes: co-inhabiting, 96; evolving natural, 79; exhausted, 80; idyllic, 69; new bio-political, 28; sacred, 36; social, 167
language, 18, 69–70, 75, 199, 202, 221, 230; plain, 160; religious, 65
Latin America, 58, 119, 222
Laudato si, 3, 10, 48, 55, 86, 88, 95–100, 105–10, 123, 182, 206, 213
law, 5, 27, 85, 99, 143, 187, 191, 205, 212; environmental, 81; intellectual property, 148; natural, 44, 50, 199; purity, x
leaders, 11, 15, 51, 118, 139, 216, 226, 229–30, 232, 235; corporate, 6, 194; imaginative, 158; local, 195; transformational, 156, 212; visionary, 158, 212
leadership: moral, 10–11, 67, 88, 91, 111, 113, 136; recent eco-social Catholic, 88; senior, 151
leisure, 43, 53–54, 100, 169, 180–85, 220; commodified, 101, 182, 185; decadent, 184; fruitful, 9, 103, 147, 160, 162, 181, 183–84, 186–87, 221; reimagining, 182
Leo XIII (Pope), 43–45
liberalism, 25, 42, 67, 100, 197; economic, 39
liberals, 86, 171; moderate, 45

liberation, 5, 57, 73, 107, 117, 124–25, 128, 131; authentic, 58; cultural, 101; theology, 43, 58
libertarian thinkers, 52, 175, 209
liberty, 129, 135, 150, 184; enduring, 5
licenses, commons-based, xi
life: animal, 76; aquatic, 192; automated, 217; cultural, 56; domestic, 54; economic, 47, 66, 92, 94, 118, 122, 160; eternal, 25; full, 25–27, 102, 117; modern, 219, 223; political, 94; post-jobs, 186; religious, 54, 183; spiritual, 49, 54, 185
lifestyles, 21, 38, 68, 74, 84, 91, 107, 110, 123, 157, 235; individual, 71; non-material, 170
life systems, planetary, 5, 37, 109, 155
limits-to-growth argument, 31, 61, 68, 70
literacy, technological, 207
living standards, 30–31, 162; higher, 179
loans, interest-free, 145
logic: countercultural, 36; moral, 174; technological, 4
Lonergan, Bernard, 129, 190, 209
love, x, 47, 53, 92, 108, 111, 129, 133–34, 215, 217, 220, 230, 233; divine, 235

M
machine learning, 3, 163, 165
machines, 6, 64, 96, 101, 141–42, 164; smart, 164, 167, 182, 215–16
management, 53, 72, 74; active, 35; adaptive ecosystem, 97; planetary, 33
Mandela, Nelson, 212
manufactured uncertainties, 16, 19, 76, 189

manufacturing, 122, 144, 163; distributed, 178; local, 144
Maritain, Jacques, 45, 56
market: liberalization, 21; mechanisms, 30, 77; transformation, 32, 38, 150, 173

markets: capitalist, 150; global, 53, 102; new, 32; open, 30
marriage, 32, 125, 158
Marxist theory, 52, 58, 127, 131
material: comfort, 185; conditions, 144; limitations, 106; progress, 31, 49, 66
maturity, spiritual, 49
McKibben, Bill, vii, 20–21, 35, 138, 155, 215–16
media, 108, 164, 168; digital, 21; social, 220
Medicare, 22
medieval, 94, 106
meditation, 117, 228
meritocracy, 21
method: ahistorical, 66; analytical, 29
methodologies, 50, 69; ahistorical, 65; contextualized, 59; contextualizing, 79; modern scientific, 95; scientific, 19; standard risk-assessment, 84
micro-grids, renewable-powered, 121, 141
middle class, 80, 169, 177, 180; global, 116; self-identifying, 104
Middle Eastern, 145
migrant workers, 222
migration, 74; forced, 80
military, 20, 24; rule, 226; spending, 59, 173
Miller, Vincent, 44–46, 87, 104–5
mindfulness, 27
minerals, 80, 141
mining, 141, 143, 163; mountaintop removal, 80
models: alternative, 142, 207; climate-science, 20;

closed-loop production, 32; community-based agricultural, 146; community-based economic, 87; fee-and-dividend, 173; hierarchical, 150; hydrocarbon-based utility, 121; innovative social-economy, 152; institutional, 40; integrated, 36; liberationist, 78; local/bio-regional, 35; post-jobs, 181; risk-management, 20; stakeholder, 109; state-socialist, 160

modernity, 6, 16–17, 21, 28–29, 86–87, 147, 157, 185–86; classical, 24; early, 16, 31, 63; late, 16–17, 24, 27; liberal, 210

modernization theory, 55, 57
monopoly, 204; current, 141
Moore, Jason, 39
Moore's Law, xi, 23, 163
morality, 72, 99, 202
movement, 18, 21, 24, 28–29, 119, 122, 124, 132–33, 155, 178–79, 184, 199–200, 227–29, 231, 235; anti-racist, 42; back-to-the-land, 47; bipartisan, 119; co-creative, 232, 234; cooperative, 122; corporate sustainability, 68; cultural environmentalist, 144; democratic, 27; divine, 234; environmental, 52; environmental-justice, 18; eugenics, 107; indigenous rights, 18; living-wage, 3; progressive, 156; revolutionary, 39; social Catholic, 45

Musk, Elon, 40, 162

N
nanotechnology, 23, 81, 189, 205

nations, 10, 51, 53, 56, 58, 61, 72, 90, 99, 116, 137, 146, 156, 173; advanced-industrial, 33, 52, 71, 74; developing, 174; impoverished, 58; industrializing, 74; poor, 57, 72, 85

natural gas, 204; fracked, 204
natural selection, 199
nature, 5–6, 30–31, 35–37, 51, 60, 63–64, 74–75, 82–83, 96–98, 106–8, 150, 198–99, 207–8, 210–11, 217–19; balance of, 69; bountiful, 69; pacific, 70; rights of, 35; romantic, 87; social, 46; systemic, 39; wild, 74

Negri, Antonio, 124, 179, 187
neoliberal: agenda, 8, 39; status quo, 171; strategy, 177
neoliberals, 29–36, 72, 77, 139, 156; economic, 32
Netflix, 2, 27
networks, 46, 90, 95–96, 153, 166; activist, 156; common digital, 163; computer, 163, 167; cooperative, 144; digital, 167; fair-trade, 153; global communications, 141; integrated global, 141; logistics, 141
nonviolence, 124; creative, 155; disruptive, 155
norms, 44, 53–54, 124, 161, 180, 194; absolute, 65; cultural, 123, 149, 169; ethical, 73; legal-formal, 92; moral, 43, 47; operative, 96; social, 94, 141; traditional, 51
nostalgia, 47, 210; rural, 69
novum, 126–27, 129–31, 133, 230
NRC (Nuclear Regulatory Commission), 190, 192–94
nuclear: annihilation, 6; arsenals, 6; incident, major, 195; power plants, 192; reactors, 190–91, 204; safety laws, 191

O
oceans, 2, 90
Octogesima, 58–61, 65–66
oil, 141, 155, 218
oppression, 57, 116, 124
order, 23, 36, 49, 54, 69–70, 72, 81–83, 110, 150, 183; ecology of, 83; global capitalist, 59; hierarchical, 82; international, 52; moral, 51, 64; natural, 83, 107; new international economic, 55, 58
organisms, 98; living, 84; synthetic, 207
organizations, 20, 142, 150–53, 156, 194; civil-society, 93, 120, 152, 156; industrial, 32; secular, 41; youth, 43
Osborne, Michael A., 162, 165–66
overconsumption, 60–61, 71
overcrowding, urban, 223
ozone: depletion, 16–17, 71; layer, 189

P
palm-oil plantations, 210
Palo Alto, 217
papal, 43–45, 48, 69, 87, 213; critique, 45; message, 73; teaching, 45, 69
paradigm: conflict, 208; shift, 36, 38;
paradox, 123, 221; social-acceleration, 117
participation, xi, 27, 40, 58, 65, 73, 77–78, 105–7, 111, 151, 153, 175, 183, 201, 221; active Catholic, 45; active Christian, 59, 110; citizen, 118; increasing, 53; political, 187; responsible, 134; rules-based, 150; voter, 148
parties, 28, 148, 174, 206–7; political, 27; responsible, 74; right-leaning, 30
Paul VI (Pope), 51, 55–58, 60, 64–66, 71, 78, 99, 131, 160, 213
peace, 42, 48, 52, 57, 74, 78, 81, 184

pensions, 137, 145; traditional guaranteed, 22
People's Climate March, 6
People's World Summit, 19
perceptions, 26, 132, 216; growing, 61; new, 60; public, 170
peril, 56, 67, 81, 96, 138
Perrow, Charles, 18
perspectives, 2, 8, 37, 41, 63, 76, 83, 88, 94–95, 111, 117, 207, 221; alternative, 55; bio-conservative, 190; critical-theoretical, 200; diverse, 88; eco-theological, 8, 201, 219; evolutionary-ecological, 79; global, 56; global eco-justice, 69, 85; green, 73; green-communitarian, 79; historical, 39; just-sustainability, 62, 70, 74, 87; late-modern, 69; liberationist, 77; philosophical, 117, 125; practice-based, 106; progressive business, 33; social-ethical, 148; socialist feminist, 52; societal, 167, 206; theocentric, 83; theological-ethical, 181
Phelps, Edmund, 148, 178
philosophers, 26, 196
philosophy, 9, 42, 82, 107; classical, 55; environmental, vii; integral-humanist, 54; mystic, 212; secular, 54, 83; social, 44
phosphorus levels, 192
photography, digital, 167
Pickett, Kate, 167, 169
Pius XII (Pope), 3, 44-45, 48–49, 51, 55, 61, 86, 185, 213, 224

planet, 2, 7–8, 18, 21, 35–37, 49, 52, 68, 78–79, 91, 95, 109, 115, 163, 170; endangered, 149; inhospitable, 138
planetary boundaries, 8–9, 63, 71, 91, 96, 140

plants: lignite coal-burning, 121;
    micro-power, 164; nuclear,
    191
platform: digital, 89; sharing, 149
pluralism, 52, 78, 153, 187; flexible
    strategic, 156
policies, 29, 31, 34, 39–40, 68, 72,
    105, 107, 109, 153–54,
    161, 173, 179–81, 208, 213;
    conflict of interest, 119;
    institutional, 91; neoliberal,
    35, 68, 137; new distributist,
    187; population-control, 52;
    sustainable-development, 74
policy implementation, 28, 197
politics, 6, 10–11, 27–29, 32–34, 90,
    94, 100, 105–6, 183, 185,
    189, 191, 194, 197, 221; late-
    modern, 197
pollution, 16, 31, 60, 95; chemical,
    18
population growth, 59, 136–37;
    rapid, 52
populations: exploited, 77;
    marginalized, 174
Populorum progressio, 51, 55, 56,
    58, 61, 71
post-carbon: economy, 87, 90, 149,
    154, 173; highly-efficient, 40;
    infrastructure, 34, 146
post-Cold War period, 69
post-jobs society, viii, 10, 40, 50,
    94, 145, 161, 172, 179–81,
    183–86, 235; inclusive, 40,
    91, 111, 136, 152, 156, 161,
    179, 181, 187; transition to,
    159, 178, 182
post-natural world, 35
post-scarcity, 170, 178
postwar decades, 168, 226
poverty, 15, 71–72, 74, 90, 157, 159,
    161, 168–71, 173–74, 177,
    180, 186, 223; endemic,
    30, 152; extreme, 169;
    persistent, 168–69
power, 22–23, 37–40, 53, 58–59,
    62–63, 80, 82, 100, 104,
    106–7, 124, 130–31, 176–77,
    206–7, 217–18; balance
    of, 137, 175; capitalist,
    187; colonial, 5; computer,
    163; distributed, 96, 158;
    divine, 211; economic, 137;
    micro-grid, 121; nuclear, 18,
    121; political, 40, 97, 148;
    transformative, 200
practices: alternative, 39; business,
    56; cognitive, 127;
    cultural, 105; deliberative-
    democratic, 28, 109, 154,
    198, 214; exploitative,
    81; interdisciplinary,
    99; manufacturing,
    141; monastic, 25; risk-
    management, 109; scientific,
    208; spiritual, 116–17,
    129–30, 183
prayer, 160, 184; contemplative, 130
predictions, ix, 34, 195; scientific, 17
prices, affordable, 144
pride, ix, 115, 226
principles, 6, 64, 66, 93, 125,
    128–29, 133, 150–52, 190,
    197, 206–7, 212; core, 78;
    ethical, 19; fundamental,
    44; operative, 64; regulative,
    207; social, 43, 45, 65, 96;
    traditional, 92
printers, 3, 89, 143–44, 146, 177
priority, 29, 38, 53, 64, 77, 94, 99,
    105, 118, 148, 154, 182–83,
    186–87, 221
prisons, private, 80
privilege, 117, 123–24
proactionaries, 28, 31, 43, 67, 85,
    206
problem solving, 152, 202
production, 7, 30–32, 37, 47,
    54, 62, 93, 144, 146–47,
    162, 168, 177, 187, 204,
    221; sustainable, 33;
    technological, 101
productivity, 3, 30, 141, 162–63,
    173, 179; high, 158;
    stagnant, 163

profits, 3, 31, 37, 47, 54, 63, 80–81, 93, 107, 144; declining, 177; maximize, 79, 218; privatize, 194; short-term, 154, 191
program: drug treatment, 174; educational, 107; federal welfare, 173; land-reform, 57; means-tested, 174; safety-net, 177; state-sponsored social, 57, 123
progress, 5, 17, 51, 75, 82, 155, 210, 212; industrial, 61, 69, 83; modern, 62; scientific, 76; self-annihilating, 197; self-endangering, 19, 81, 189; social, 51, 56; technological, 34; techno-economic, 7, 10, 27–28, 61, 64, 73, 82, 196, 210; techno-scientific, 70
project: alternative social, 67; church-related social-action, 41; infrastructure, 155; social-democratic, 138; technological utopian, 213
property, 44, 53, 80, 155, 176–77, 195; intellectual, 143; private, 52–53, 77, 174, 176; rural, 148
prophets, ix–x, 116
protection, environmental, 34, 37, 39, 75, 81, 171
Protestant work ethic, 4
protests, 6, 68, 194; anti-globalization, 18; nonviolent, 73; utopian, 131
provincialism, 101
prudence, 35, 48, 82, 85
psychology, positive, 94

Q
Al-Qaeda, 20
Québec, 68, 119–20, 152
questions: basic, 16, 39, 168, 208; disquieting, 143; ethical, 182; evaluating, 177; hard, 196; practical, 146, 221; social, 8, 15, 43–44, 51, 64, 223

R
race, 7, 163, 231; superpower arms, 50
radical reform, 40, 55, 111, 149, 152, 154, 156, 171, 180, 218; politics of, 58, 155
radioactive contamination, 18, 192
ratepayers, 191, 204
rationality: critical, 210; practical, 208
recommendations, 194, 207; strategic, 80
Redemptor hominis, 62, 73
reforms, 37–38, 44, 48, 72, 94, 103, 105, 109, 154–55, 213–14; agrarian, 78; eco-modern, 38; major, 87, 154; social, 45, 72; social-democratic, 226
region, 57, 79–81, 146, 154, 193, 202, 204; backward, 5; coastal, 193; developing, 121; exploited, 79; local, 118; poorest, 118; rural-wild, 78
regulation, 17, 28, 207–8; environmental, 31, 37
regulatory: infrastructure, 207; regimes, 72
relationships, ix–x, 50, 75, 92, 96, 100, 108, 111, 126, 129, 182, 201–2, 220; dialogical, 36; evolving, 91; humanity-nature, 97, 207; intimate, 146, 160; parent-child, 204; social, 51; sustaining, 46
religion, x, 25, 71–72, 107, 127, 129–30
renewal: cultural, 10, 28, 94; democratic, 98, 117, 122, 125, 128, 131, 136, 142, 153, 156
research, vii, 61, 91, 94, 115, 165, 170, 173, 207–8; historical, 25; ongoing, 95, 200; scientific, 83, 90, 100; stem cell, 32
resources, 4, 26–27, 30–31, 40, 51, 61, 63, 71, 83, 167, 176, 192,

214; cultural, 197; economic, 40; educational, 22; material, 141; natural, 31, 64, 72, 74, 141, 222; nonrenewable, 71; non-renewable, 74; wealth-generating, 176
responsibility, viii, 46, 49, 54, 60–61, 63, 70–71, 74, 85, 92, 99, 132, 196, 199, 204–6; elite, 73; moral, 201; sacred, 197
revolution, 45, 131, 147, 218
Rifkin, Jeremy, 34, 141, 163, 168
rights: basic, 18; economic/social, 52–53; environmental, 34; individual, 44, 175; intellectual, 149; non-transferable, 151; protecting, 53, 78; property, 30; reproductive, 52; universal, 66
risk management, 17, 19, 86, 196
risks, 16–19, 21, 34, 76, 82, 84–86, 136, 138–39, 162, 166, 177, 181, 189, 193–96, 204–6; calculated, 61; long-term, 76; nuclear, 195
robotics, x, 3, 23, 81, 165–66, 172, 189, 205, 219–20, 233; advanced, 163
roots, 64, 93, 99, 133, 180, 194, 202, 212; religious, 106; systemic, 37
Rosa, Hartmut, 9

S
safety, 19, 84, 103, 189, 193–94, 216
scarcities, 31, 186
schools, 94, 156, 217; cultural environmentalist, 149
Schumacher, E. F., 35, 47
science, 17, 34, 36, 82, 85, 96, 100, 106, 127, 129–30, 135, 178, 183, 205–6, 208; applied, 85; modern, 42, 54, 66; social, 79, 96; subversive, 85
scientism, 55, 83, 106

scientists, 6, 8, 29, 49, 162, 203; atomic, 6; social, 149
scripture, 65–66, 75, 88
sea levels, rising, 190, 193, 203
secular utopias, 131
security, 25, 100, 123, 131, 142, 161, 167, 180, 210; economic, 144, 172, 181; national, 146; social, 22, 172, 177
self-interests, enlightened, 56, 78, 98, 143
self-transcendence, 107, 129, 160
Seligman, Martin, 26, 135
sensitivity, 133; ethical, 186; mutual, 229
sentience, 198, 200; embedded, 203
services, 33, 35, 50, 53–54, 63, 101, 105, 120, 127, 130, 141, 144, 147–48, 150–51, 154
shock, 23, 42, 139, 156; future, 171; present, 4
silence, 116, 123, 190, 194, 197; culture of, 123
simplicity, 35; voluntary, 186
sin, 66, 73; basic, 129; personal, 64, 72; social, 62, 64, 66, 72, 77
slave labor, 74, 182
slavery, 49, 185; wage, 46, 89, 186
social analyses, radical, 37, 57, 59
social change, 4, 11, 23, 28, 35, 56–58, 66, 77, 124, 127, 136, 157, 160; progressive, 52; radical, 78; rapid, 123
social empowerment, 117, 122, 125–26, 128, 131, 136, 142
social injustice, 63–65, 87, 102, 117
social learning, 120, 140, 208, 229
social teaching, Catholic, 13, 45–46, 59, 73, 87
social theories: contemporary, 9, 65; liberal, 52, 59; modern, 65; organic, 50, 57; radical, 50, 52–53, 66, 73
social thought, 8–9, 42–43, 45, 53, 94, 111, 117, 131, 160
socialism, state, 46, 77
society, 9–10, 33–34, 44–45, 47–48, 52–53, 96–97, 99–100,

130–31, 156–57, 167, 169–72, 175, 180–81, 185, 209–13; communitarian, 47; consumer, 123; democratic, 176; divided, 180; emerging, 182; laboring, 185; medieval, 47; modern, 25–27, 42–43, 48, 53, 55, 64, 101, 147, 210; new, 58, 128; post-capitalist, 36
socioeconomic inequalities, 99, 171; widening, 68, 90
soils, 97, 133, 218; degraded, 74
solar, 204
solidarity, 35, 64, 72–74, 77–78, 94, 128, 155, 177; global, 56; limited, 210; peasant, 18; universal, 65, 72
Sollicitudo rei socialis, 71–73
solutions, eco-fascist, 197
space, 2, 93, 95, 108–9, 140, 171, 174, 195; deep, 133; defined, 98; green, 18; political, 126
Spain, 121–22, 222
species, vii, 5, 9, 72, 97–98, 127, 138; endangered, 106
stability, 137, 154, 178, 200; dynamic, 24; psychic, 128; social, 39; telic, 83
stakeholders, 109–10, 120, 195, 206, 208
standards, strong environmental, 32, 84, 154
stewardship, 63, 65, 70–72, 77, 95, 206
strategy, 30, 32–33, 38, 40–41, 45, 58, 62, 87, 146, 150, 153–56, 158, 161, 171, 177–79
structures, 73–74, 93, 104–5, 117, 123–25, 142, 152, 181, 194, 198, 200, 202, 229–30, 234; dysfunctional, 72; hierarchical, 151; public, 108; self-governing, 90; stable, 123
subsidiarity, principle of, 44, 46, 53, 77, 94, 97–98, 151
Suhard, Emmanuel (Cardinal), 2

suicide, 74, 167
superdevelopment, 71, 73
sustainability, 18, 29, 32, 34, 70, 78, 202, 234; environmental, 117, 131, 136
sustainability science, 8, 36, 95–96, 98–99, 200
sustainable: abundance, vii–viii, 8–9, 106, 116, 118, 131–32, 136, 140, 145, 156–57, 161, 178–79, 187, 190, 233; communities, 10, 60, 80–81, 91, 99, 111, 149, 153, 202–3; development, 9–10, 29, 32, 41–43, 68, 71, 74, 79, 81, 87, 90, 98, 109, 139, 208; growth, 30–32, 39; society, 29, 38, 170
sustainable-development debate, 67, 69–70, 73, 78, 87
system, 38, 40, 46, 96, 98, 117, 123, 131, 138–39, 150–52; accounting, 143; adaptive, 150; automated, 141, 143; closed-loop, 191; exploitative, 59, 64; financial, 137; international, 72; renewable energy, 98, 121; wage, 52, 179
systemic change, 40, 62, 67, 77, 79, 98, 117, 122, 124, 135–36, 153, 155, 180, 234
systemic risks, 9, 16, 136, 138–39, 157, 161

T
Taleb, Nassim, 7, 20, 76, 86, 126, 195–96
tango, xi, 215, 221–26, 229–32, 235
tariffs, 32, 121, 174
taxes, 173; carbon, 32, 149, 154, 171, 173–74
technocracy, 48, 83, 85–86, 100, 106–7
technocratic paradigm, dominant, 10, 55, 99–100
technological: development, 54, 81, 83, 106, 163, 198–99;

innovation, 3, 23, 25, 28, 30, 35, 37, 85, 97, 101; society, 9, 53, 102, 184; unemployment, xi, 10, 22, 50, 137, 161–63, 166–68, 172–73, 182, 186
technology, 3, 9–11, 23–25, 48–49, 51–55, 63–64, 82–83, 100–108, 163–64, 188–91, 195–99, 203–7, 213–14, 218–21, 234–35; emerging, 3, 6, 10, 17, 23, 29, 43, 81, 85, 94, 141, 161, 189, 205–6; modern, 48–49, 64, 100–101, 103, 106, 186, 220; new, 15–17, 32, 34, 67, 85, 100, 162, 172, 190, 203, 205, 214; post-carbon, 216; renewable, 204; solar, 164
theology, 9, 55, 65, 82, 130–31, 133, 184; medieval, 25; natural, 55, 69; radical, 234; scriptural, 66; traditional, 70
theory: contemporary ecological, 83; dependency, 57–59, 66; ecological-modernization, 33; evolutionary, 94, 122; false economic, 64; modern moral, 103; modern risk-management, 85; orthodox economic, 52; social-acceleration, 207; world systems, 57, 87
Third World, 50, 71
Thomism, 45
Thomist metaphysics, 82, 110
threats, 4, 48, 60, 62, 85, 142, 189, 197, 207; irreversible, 205; power of, 18, 62
thresholds, critical, 157
Toffler, Alvin, 23, 171
tools, 46, 54, 105, 146, 199, 207; premodern, 146; productive, 147; value-neutral, 100
tradition, xi, 10, 21, 45, 60, 87–88, 92, 95, 104, 116, 147, 202, 221, 235; ancient Christian, 75; aristocratic, 42; cultural, 79; distributist, 147; humanist, 102; kataphatic-mystical, 110; liberal democratic, 100; premodern, 69; romantic, 74; social, vii, 97; social-encyclical, 43; theological, 219

trajectories, 56, 126, 152, 201; current, 195; dangerous, 209; potential, 214; strategic, 153
transcendence, 25, 111, 128, 209–10, 212
transformation, 34, 97, 149, 202, 231; social, 78, 94, 125, 140, 152, 157–58
transition, 5, 7, 10, 38, 40, 99, 143, 146, 149, 152, 154, 170, 172, 178–80, 182; historical, 16; rapid, 90, 98
transparency, 92, 143, 189, 193, 195, 203–4
Trump administration, 171, 173
trust, 92, 229
truth, 77, 92, 211, 232; anthropological, 66, 211–12; inconvenient, 2, 10, 19, 159, 186; religious, 56; salvific, 54; soteriological, 211–12
Tubman, Harriet, 212
Turkey Point Nuclear Generating Station, 190–97, 204
twenty-first century, ix, 10, 15, 34, 142, 144, 146, 177; early, 136
tyranny, 2, 7, 136

U
uncertainty, 7, 10, 17, 19, 34, 82, 84, 91, 195, 203
underdevelopment, 51, 57–58, 71–72
underemployment, 164, 186
unemployment, 24, 45, 119, 152, 162, 164, 167, 171, 177, 186; chronic, 80; high, 167, 171; rising, 167; structural, 3, 139, 180
unions, 21, 119, 177

urbanization, 39, 59, 82, 97, 136
utility, 47, 77, 103, 121, 149, 191, 204
utopia, 169, 174; concrete, 58, 81, 125–27

V
values, 56–57, 63, 66, 95, 97–98, 130–31, 140, 143, 147, 150, 174, 176, 185–86, 209–10, 212–13; core, 52; democratic, 155; environmental, 37; natural, 149; spiritual, 54; therapeutic, 74
Vatican II, 50
violence, 37, 123, 175, 202; institutionalized, 57
vision, 34, 45, 51, 63, 81, 95, 122, 135, 140, 150, 181, 212, 232; alternative, 7, 116, 153; arcadian, 70; church's, 71; cosmopolitan, 178; green-communitarian, 78; medieval millenarian, 127; moral, 176; prophetic-utopian, 130–31; sacramental, 75; social-ethical, 43, 46, 66; teleological, 219; theological, 110; utopian, 127
vocation, 49, 134, 136, 160, 234; human, 47, 52, 63, 79, 111, 234; transformative, 63
vulnerabilities, 7, 18, 45, 97, 122, 124, 138, 142, 172, 189, 193–94, 203

W
wage labor, 94, 180–82, 185–86
wages, 44, 122, 174
war, 37, 48, 83
waste, 61, 71
water, 35, 39, 61, 133, 146, 157, 171, 176, 191
wealth, 44, 46, 48, 51, 54, 90, 94, 106, 142, 146, 165, 167, 169–70, 174, 176–77; democratize, 81, 90, 118, 144
weather events, 139, 157, 193
Weeks, Kathi, 125, 172, 179
WEF (World Economic Forum), 39, 139
welfare, 210; means-tested, 176
Wilkinson, Richard, 167, 169
wisdom, 48, 51, 56, 82, 147; social, 10, 94, 147, 160, 183; traditional, 36
women, 63, 88, 145, 185, 223, 225, 228
work, 47–49, 53–55, 62–63, 65, 125–26, 151–53, 159, 161, 168–70, 172, 174–75, 178–83, 185–87, 217–18, 222; professional, 54, 183
workers, 53–54, 62–64, 66, 68, 101, 162–63, 165–66, 168, 170, 172, 174–78, 180, 183, 185, 194
work ethic, 123, 172, 174, 180, 182–83, 185–86, 235
workplaces, 63, 121, 171, 179
world risk society, 5–7, 10, 15–19, 22, 29–30, 35, 38, 42, 59–60, 66, 76, 81, 86, 189–90, 196; emergence of, 17, 62, 79; emerging, 40, 43, 50, 59–60
worldview, 198–99; evolutionary-ecological, 200; materialistic, 83
World War I, 20, 223
World War II, 189
World Wide Web, 143

Z
zones, 91, 149; buffer, 148; high-risk danger, 91; plume exposure pathway, 192; rural, 149

www.ingramcontent.com/pod-product-compliance
Lightning Source LLC
Chambersburg PA
CBHW030613230426
43661CB00053B/1971